STUDY GUIDE

for use with

HÉLÈNE I. SAVARD
School of Environmental & Natural Resource Sciences
Sir Sandford Fleming College

NELSON
━━━━━✦━━━━━ ™
THOMSON LEARNING

Australia • Canada • Mexico • Singapore • Spain • United Kingdom • United States

NELSON

THOMSON LEARNING

™

Study Guide to Accompany
Our Environment: A Canadian Perspective
Second Edition

by Dianne Draper
Prepared by Hélène I. Savard

Editorial Director and Publisher:
Evelyn Veitch

Acquisitions Editors:
Kelly Torrance, Edward Ikeda

Marketing Manager:
Janet Piper

Developmental Editor:
Elke Price

Production Coordinator:
Hedy Sellers

Creative Director:
Angela Cluer

Cover Design:
Katherine Strain

Cover Image:
© Barrett & MacKay
Photography Inc.

Printer:
Webcom Limited

FOREWORD

Although there are cumulative anthropogenic impacts on Earth, studying contemporary science may help us better understand ecosystems, how they work, and our place in them. Ecological and environmental information, emerging skills and technologies, combined with positive attitudes, may allow us to "fix" some of the negative impacts and thereby attain a sustainable future.

In *Our Environment*: *A Canadian Perspective, Second Edition,* you will find detailed examples of current ecological and environmental applications. This is a timely text of Canadian and international importance!

The Study Guide was designed to *help you, the learner,* understand and appreciate the content of the textbook. Perhaps you may choose to work in teams. Each chapter follows Dr. Draper's textbook—*Our Environment: A Canadian Perspective, Second Edition.*

Each chapter contains *Learning Objectives*, *Key Terms* with page references, *Notes,* and *Tracking Your Progress Questions*. Use *Our Environment: A Canadian Perspective* to help you learn the *Key Terms*, and to fill in the *Notes* section prior to attending corresponding sessions. You may find the section *Tracking Your Progress* section useful as it contains multiple choice questions to test your progress. However, be aware that not every important issue discussed in the text chapter is covered in these questions. It is strongly recommended that you also complete the *Chapter Questions* found at the end of each textbook chapter. Another valuable learning tool is to look up some of the book and Web references found at the end of the textbook chapters and to attend class or laboratory sessions when offered.

I gratefully acknowledge Dianne Draper, author of *Our Environment*: *A Canadian Perspective, Second Edition*, and Elke Price for her kind help in editing this work.

Hélène

Ecosystem Management Professor
School of Environmental & Resource Sciences
Sir Sandford Fleming College
Lindsay, Ontario
K9V 5E6

hsavard@flemingc.on.ca
http://www.flemingc.on.ca/Programs/Natres/

USING THE STUDY GUIDE

The *Study Guide* is designed to accompany *Our Environment: A Canadian Perspective, Second Edition* by Dianne Draper. The headings, and intended uses of the sections, are as follows:

STUDENT LEARNING OBJECTIVES

You want to know what is *expected* to be learned. Make your learning easier by checking off the *Student Learning Objectives* as they are mastered. Increase your depths of understanding by researching and reading some of the *References* listed at the end of each *textbook* chapter. The time invested on these items will be well spent as they encompass the main ideas of each chapter.

KEY TERMS

It is important that you understand and adopt the vocabulary of ecology and environmental science. Research the *Key Terms* as they arise, and then continue to use them throughout the year so that you learn the words by immersion. Focus your attention on the *Key Terms* by making flash cards. Write the term on one side, and the meaning, with an example, on the backside. By applying examples to each one, you will remember their context. To make your learning even easier, review your flash cards and base your discussions on *Key Terms*. These basic exercises may facilitate your use of *Key Terms* in discussions, interviews, and on the job. Finally, test your comprehension of the *Key Terms*, using questions of various formats, such as fill in the blanks, true or false, multiple choice, crossword puzzles, or define these terms. Each individual can be assigned a task to help build each team-member's comprehension skills.

NOTES

The *Notes* section is intended to suggest the main ideas and themes that may be presented by lecturers covering the material presented in this *textbook*. Read the *textbook* and fill-in the appropriate *Notes* section prior to attending the corresponding lecture. By going to class prepared, you will be ready to add complementary information to your notes and to ask for clarification.

TRACKING YOUR PROGRESS

Each *Study Guide* chapter ends with *Tracking Your Progress*, a multiple-choice format section intended to help the learner review important points discussed in the *textbook*. Each question ends with a *textbook* page reference. This page reference indicates the location of each answer. Please treat this section with caution as not all important points are covered here. You will be more thorough by also attempting the *Chapter Questions* found at the end of each chapter in the textbook and the Chapter Quizzes on the book Web site:

www.environment.nelson.com

USING THE NELSON THOMSON LEARNING WEB SITE

The Nelson Thomsom Learning site www.environment.nelson.com is designed to complement *Our Environment: A Canadian Perspective, Second Edition*. Some of the site's features include:

- text updates

- chapter quizzes

- environmental topics and newspaper articles with Web links

- Web links on a chapter-by-chapter basis

- comments and questions

- glossary

- about the author

- about the book

You can make good use of the Web site by accessing the URL links which are listed on a chapter-by-chapter basis. You can assess your understanding of the textbook, and prepare for your tests by trying the online Chapter Quizzes. Visit other college and university Web sites for information regarding various programs and job opportunities. For example try: *http://www.flemingc.on.ca/Programs/Natres/*

Have fun as you learn!

CONTENTS

CHAPTER 1—OUR ENVIRONMENT: PROBLEMS AND CHALLENGES
(pp. 2-25)

STUDENT LEARNING OBJECTIVES

After studying this chapter you should be able to:
- identify a range of local, regional, and international environmental issues of relevance to Canadians and all of Earth's citizens
- appreciate the ways humans are linked with Earth's ecosystems
- describe the root causes of environmental problems
- discuss the concepts of sustainable development and sustainability
- identify and summarize the guiding principles of sustainability
- define and discuss all of the KEY TERMS listed below

KEY TERMS (defined at the page number shown and in the glossary)

biodiversity	14	exponential growth	3
biosphere	9	natural capital	12
carrying capacity	14, 16	non-renewable resources	12
country food	5, 20	renewable resources	12
ecological footprint	19	stewardship	23
ecosphere	5	sustainability	3, 14
ecosystem	5	sustainable development	14
environment	5		

NOTES

Introduction
Canadians know they inhabit some special places on a beautiful planet.

The changing global environment
*"...human activities have increased the pace of **environmental** change and have had dramatic impacts on the quality and productivity of the planet's **ecosystems**."*

Ecosystem	*Environment*

Linkages: Humans as part of ecosystems

Major causes of environmental problems
World scientists' warning to humanity (see Enviro-Focus 1)

Human population growth

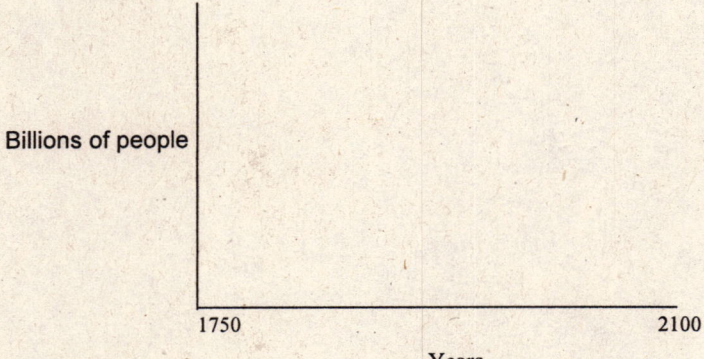

Billions of people

1750   2100

Years

Over-consumption of resources and natural systems

"Humans have defined resources as anything useful that serves our needs and is available at a price we are willing to pay."

Renewable resources	Non-renewable resources

Pollution problems: *Historically, humans have discharged their wastes without due regard for the long-term and cumulative impacts on the environment. Change of attitudes is a goal.*

Related themes: *What kind of standard of living do we want for our descendants?*

Sustainability challenges

"Sustainability is a major theme and challenge related directly to the future of this planet."

Sustainability	Biodiversity	Sustainable development

Our Common Future (WCED 1987) and carrying capacity (WCU 1991)

Guiding principles of sustainability

Ecological sustainability (see also carrying capacity)

Ecological footprints

An ecosystem approach

Social sustainability

Economic sustainability

The Precautionary Principle

Environmental stewardship

Monitoring for sustainability

Toward sustainability

TRACKING YOUR PROGRESS
Chapter 1—Problems and Challenges

Multiple Choice Questions:

1. Exponential growth in resource use and population has caused:
   a) disappearance of grasslands, wetlands and forests
   b) poisoning of oceans and water bodies
   c) top soil erosion
   d) extinction of wildlife species
   e) all of the above
(p. 3)

2. Canadian media have carried stories about:
   a) the growing ice pack
   b) illness and death from contaminated juice supplies
   c) fish to produce human protein for hair replacement
   d) collapse of cod and salmon fisheries
   e) (b) and (c) are both correct
(p. 4)

3. Polychlorinated biphenyls (PCBs) have been found in:
   a) granite
   b) Arctic country food (local meat and fish)
   c) Lodgepole pines
   d) Antarctic rock
   e) none of the above
(p. 5)

4. Ecosystems are affected by human activities and have:
   a) assimilative and infinite productive capacities
   b) assessing and finite productive capacities
   c) assertive and infinite productive capacities
   d) assimilative and finished productive capacities
   e) assimilative and finite productive capacities
(p. 5)

5. The World scientists who signed the Warning to Humanity (1993):
   a) blamed the United States and Canada
   b) represent the views of a few deep ecologists
   c) lied about the negative impact of humans on our ecosystems
   d) represent the world view of more than 1670 scientists from 71 countries
   e) stated that the Earth was finally in good repair and thanked everyone on Earth
(Pp. 6-7)

6. A suggestion from the World Scientists' Warning to Humanity was that:
    a) the irreversible loss of species is very serious
    b) the current moderate decline of some species is unexplainable
    c) with cloning techniques we can now reverse the trend of extinction
    d) overall destructive pressure on species has been very low
    e) exploitation of species endangers our water systems
(Pp. 7-9)

7. The World Scientists' Warning to Humanity suggests that even at this moment:
    a) 1 person in five lives in absolute poverty
    b) 1 person in fifty lives in absolute poverty
    c) 1 person in five hundred lives in absolute poverty
    d) 1 person in five thousand lives in absolute poverty
    e) poverty has finally been eradicated
(Pp. 7-9)

8. If vast human misery is to be avoided and this planet is NOT to be irretrievably mutilated:
    a) stewardship is required
    b) move away from fossil fuels
    c) halt deforestation
    d) eliminate sexual equality, and never guarantee women control over their own reproductive decisions
    e) (a), (b) and (c) are all correct
(Pp. 7-9)

9. The Warning suggests that we need help of many, including:
    a) world's religious leaders
    b) world's business and industrial leaders
    c) scientists – natural, social, economic and political
    d) (b) and (c) only
    e) all of the above
(Pp. 7-9)

10. By 1999 the human population had reached about:
    a) 10 billion
    b) 9 billion
    c) 6 billion
    d) 2 billion
    e) 1.3 million
(Pp. 10-11)

11. United Nations statistics indicate that each day almost _____ people die from starvation or related illnesses
    a) 11
    b) 110
    c) 1100
    d) 11 000
    e) 110 000
(Pp. 10-11)

12. Humans can be viewed as being part of what systems or worlds?
    a) natural systems
    b) biological systems
    c) sociocultural systems
    d) not part of ecosystems
    e) (a), (b), and (c)
(Pp. 10-11)

13. An anthropocentric view is always a:
    a) correct viewpoint
    b) human viewpoint
    c) ecosystem viewpoint
    d) species based viewpoint
    e) resource based viewpoint
(Pp. 10-11)

14. Growth in total world population suggests that by 2050 the total population will be
    _____.
    a) 8.9 billion
    b) 20.6 billion
    c) in a moderate decline thanks to UN initiatives
    d) in no more poverty
    e) extinct
(p. 11)

15. The following are examples of renewable natural capital if harvested at acceptable rates:
    a) deer, gas, limestone, water
    b) insects, trees, wind energy, mammals
    c) forests, solar energy, gravel, fish
    d) whales, crawfish, salmon, fossil fuels
    e) petroleum, hydrogen gas, electricity, batteries
(p. 12)

16. Energy and material consumption in Canada is _____ the world's average.
    a) equal to
    b) 3 to 5 times
    c) 30 to 50 times
    d) any of the above
    e) none of the above
(p. 12)

17. International awareness of the pollution problem was promoted by the:
    a) 2001 oil spill from the tanker *Jessica* off the Galapagos Islands
    b) 2000 Esmeralda cyanide mining spill into the Danube River
    c) 1995 tailings dam break at the Omai gold mine in Guyana
    d) (a) and (c) only
    e) all of the above
(Pp. 12-13)

18. Canadians withdrawal what percentage of the global resources' fresh water?
    a) 3
    b) 15
    c) 35
    d) 60
    e) 99.9
(p. 12)

19. Kaufmann et al. (1994) said that sustainability refers to the ability of an ecosystem to maintain ecological processes and functions, biodiversity and productivity over:
    a) time
    b) space
    c) ecosystems
    d) humanity
    e) Earth
(p. 14)

20. To be sustainable the World Conservation Strategy (1980) stated that development must:
    a) be insensitive to short- and long-term alternatives
    b) take social, economic, and ecological factors into account
    c) include living and nonliving resources
    d) all of the above
    e) (b) and (c) only
(p. 14)

21. From the Bruntland Report (Our Common Future, 1987) the popular usage of sustainable development became an oxymoron, but the World Conservation Union et al. (1991) says it:
    a) calls for ecological and social transformation to a world of stability and justice
    b) means more growth, or more sensitive growth
    c) is a reformed version of the status quo
    d) is progressive development
    e) leads to improvement in the quality of life while living within the carrying capacity of supporting ecosystems

(p. 14)

22. Sustainable objectives identified by the Canadian federal government:
    a) sustain our natural capital
    b) protect health of Canadians and ecosystems
    c) focus more on current political agendas
    d) promote capitalism world wide
    e) (a) and (b)

(Pp. 16-18)

23. The ecological footprint analysis has shown two conflicting trends, the truth is:
    a) Earth shares are rising
    b) the ecological footprint for people in the Lower Fraser Valley, BC is estimated at 19 times the land area they occupy
    c) local carrying capacity will increase when resources are imported
    d) the biological optimum has been reached
    e) ecological footprints are shrinking while Earth shares are growing

(Pp. 18-19)

24. Economic sustainability must emphasize:
    a) ecological benefits
    b) jobs and good income
    c) jobs and ecological benefits
    d) jobs, good income, and ecological benefits
    e) jobs, good income, and economical benefits

(p. 23)

25. Canada and many countries lack adequate _____ and _____ regarding enviro-change.
    a) jobs, income
    b) time, money
    c) food, energy
    d) space, housing
    e) records, data

(p. 23)

26. When the lack of full scientific certainty is used as a reason for postponing cost-effective measures to prevent environmental deterioration:
    a) the Precautionary principle is demonstrated
    b) sustainable development occurs
    c) stewardship spreads
    d) less impact occurs over time
    e) none of the above

(p. 23)

27. Lerner (1993) described stewardship as:
    a) active earthkeeping
    b) wealth
    c) rehabilitation
    d) mitigation
    e) recycling

(p. 23)

28. What did Toronto residents do to demonstrate stewardship?
    a) took the metro
    b) hitchhiked
    c) cleaned up the Don River
    d) incinerated their household trash
    e) recycled their cars

(p. 23)

CHAPTER 2—ENVIRONMENTAL STUDIES: SCIENCE, WORLDVIEWS, AND ETHICS
(pp. 26-48)

STUDENT LEARNING OBJECTIVES

After studying this chapter you should be able to:
- appreciate why a scientific understanding of our environment is important in making decisions about sustainability
- outline the basic methods of science
- identify the different values and worldviews present in environmental controversies
- discuss the key historical approaches to conservation as well as current approaches to environmentalism
- understand the practical and moral reasons for valuing environments
- describe the concerns associated with environmental ethics
- define and discuss all of the KEY TERMS listed below

KEY TERMS (defined at the page number shown and in the glossary)

NOTES

Introduction
This is about the historical and the current roles and influences of science, worldview, and ethics.

Science and the environment
Why should we want to think scientifically about the environment?

What is Science? *Scientific views, such as Darwin's theory of evolution, can conflict with or complement generally accepted social benefits.*

Assumptions in science: *Botkin & Keller (1995) suggested five basic assumptions made by scientists*

Thinking scientifically
Scientists make use of two kinds of reasoning

Deductive reasoning (thinking)	Inductive reasoning (generalization)

Scientific Measurement: *The drowning of 9600 Caribou in northern Quebec in 1984 is thought to have resulted from increased flow in the Caniapiscau River, due in part to the operation of dam systems. Why would one use statistical significance in this case?*

The methods of science

Misunderstandings about science

Use of language: *Scientists may disprove things, they cannot establish absolute proof. The term "proven scientifically" is inaccurate use of scientific language.*

Value-free science: *Is the use of bananas for immunizing developing nation children against life-threatening diseases value-free?*

The scientific method

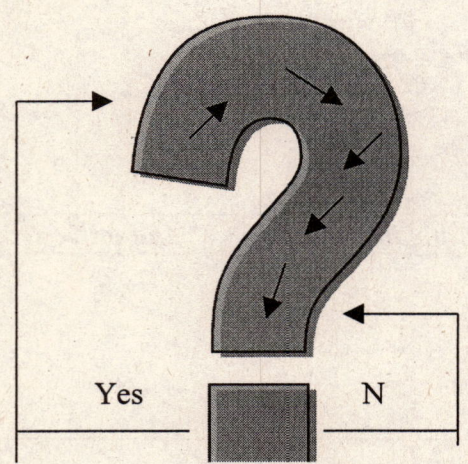

Feedback processes in scientific investigation (see Figure 2-1)

Complexity, values, and worldviews: *An insistence that we fully understand a problem before taking action may lead to "paralysis by analysis".*

Science and environmental decision making: *Environmental decision making should incorporate all possible worldviews and values.*

Worldviews and values

Worldviews are "sets of commonly shared, values, ideas, and images concerning the nature of reality and the role of humanity within it".

Expansionist and ecological worldviews *(see also Table 2-2)*

Expansionist Worldview	Ecological Worldview

Conservation in the early 20th-century

Wise management	Righteous management

Environmentalism

Environmental values and ethics

Environmental values	Environmental ethics

TRACKING YOUR PROGRESS
Chapter 2—Environmental Studies: Science, Worldviews, and Ethics

Multiple Choice Questions

1. However remote the environment might seem to urban Canadians, it continues to play a vital role in:
   a) everyday lives
   b) rural Canada
   c) how people have changed
   d) our sustainable present
   e) our demise
   (p. 27)

2. Which of the following is the most precise measurement?
   a) 4.0 kg
   b) 4.01 kg
   c) 4.001 kg
   d) impossible to tell
   e) (a), (b), and (c) are all the most precise
   (Pp. 29-30)

3. In an experiment where "Mow-chow" food increased the weight of cats, the cats' weight is:
   a) the denominator
   b) the numerator
   c) the dependent variable
   d) the independent variable
   e) the control
   (p. 31)

4. Which of the following is an example of qualitative data?
   a) pumpkin size measured as small, medium, or large
   b) length of fish measured in millimetres
   c) size of your shirt: large, medium or small
   d) speed of germination measured in days
   e) number of wing beats per second
   (p. 31)

5. Prior to establishing a conclusion from the data, the scientist will:
   a) draw an inference
   b) accept the hypothesis
   c) develop a question
   d) interpret the data
   e) reject the null hypothesis
(p. 33)

6. Postponing action on global warming until all causes are found may lead to:
   a) expert understanding of the issue
   b) paralysis by analysis
   c) a clear idea of cause and effect
   d) precautionary principle
   e) more ozone
(p. 35)

7. Aldo Leopold and Rachel Carson were:
   a) righteous management conservationists
   b) wise management expansionists
   c) promoted sustainable exploitation
   d) expansionists
   e) none of the above
(p. 37)

8. Which of the following is considered to be part of the Expansionist worldview?
   a) the unity of human life with nature
   b) the kinship and relatedness of all life forms small and large
   c) quality may be more important than quantity
   d) human activities must work within the limitations of the planet's ecosystems
   e) the accumulation of personal wealth through the exploitation of nature
(Pp. 37-39)

9. The notion that all organisms and entities in the ecosphere have intrinsic worth is:
   a) biocentric equality
   b) deep-ecology
   c) self-realization
   d) all of the above
   e) none of the above
(p. 43)

10. A perspective that explains the Earth in terms of a self-regulating process came from:
    a) Aldo Leopold
    b) UNESCO
    c) Ecofeminists
    d) James Lovelock's Gaia hypothesis
    e) WWC
(p. 43)

11. The organization that hopes to develop practical strategies to sustainable water use is:
    a) Conservation Authority of the World
    b) World Water Council
    c) United Nations
    d) Ministry of the Environment
    e) World Water Vision Project
(Pp. 42-43)

12. The fact that the gut of a Howler monkey provides a function in plant germination is:
    a) a moral justification for conservation
    b) an aesthetic arguments for conservation
    c) an utilitarian justification for conservation
    d) an ecological justification for conservation
    e) an excuse for conservation
(p. 45)

13. Which view best supports the notion that parakeets should be conserved because of their beauty?
    a) the moral justification for conservation of nature
    b) the aesthetic argument for the conservation and protection of nature
    c) the ecological justification for conserving nature
    d) the utilitarian justification for the conservation of nature
    e) the economical justification for minimal conservation
(p. 45)

14. A land ethic that suggests that each one of us is a steward of our environment came from:
    a) Rachel Carson
    b) Aldo Leopold
    c) Gifford Pinchot
    d) Clifford Sifton
    e) Dianne Draper
(p. 46)

CHAPTER 3—EARTH'S LIFE-SUPPORT SYSTEMS
(Pp. 50-89)

STUDENT LEARNING OBJECTIVES

After studying this chapter you should be able to:
- identify and describe the Earth's major components
- outline the components and structure of ecosystems
- discuss ecosystem functions and their interconnections
- explain how ecosystem population dynamics work
- identify the major forces of change and adaptation affecting the Earth
- identify the key features of living systems that help humans learn to live sustainably
- define and discuss all of the KEY TERMS listed below

KEY TERMS (defined at the page number shown and in the glossary)

NOTES

Introduction

" All life exits in a thin layer wrapped around the globe, caught between the molten heat of the earth's interior and the cold immensities of space. The biosphere, the only part of the entire universe known to support life – proportionately is no thicker that the shine on a billiard ball." (Lean and Hinrichsen, 1992)

Matter *is the material of which things are made, the stuff of life*

Matter Quality

High-quality matter	Low-quality matter

Energy *is the ability or capacity to do work, is what enables us to move matter from one place to another or to change matter from one form to another.*

Energy quality

High-quality energy	Low-quality energy

Physical and chemical changes in matter

Physical change	Chemical change

The law of conservation of matter

First and second law of energy

First law of thermodynamics	Second law of thermodynamics

Earth's Life-Support Systems

Earth's Major Components

Connections on Earth: Life on Earth depends on three pervasive & interconnected factors.

Ecology (*see also Levels of Biological Organization, Figure 3-4*)

Examples of abiotic, biotic, and human cultural components

Abiotic component	Biotic component	Human cultural component

Biodiversity

Genetic diversity	Species diversity	Ecological diversity	Human cultural diversity

Earth's Major Biomes (Figure 3-5)

Ecozones of Canada (Figure 3-6)

Types of Organisms *(see pages 61-63)*

Prokaryotic	Eukaryotic				
				Animalia	
	Protista	Fungi	Plantae	Invertebrates	Vertebrates

Components and structure of ecosystems

*All Ecosystems, large and small are made up three main features: **"Components"** (biotic and abiotic parts) which give **"structure" and** serve special **"functions" (processes)** on Earth. This table looks at a few examples as they may apply to a tree.*

Component of a tree	Structure of a tree	Function or process of a tree
• Roots • Trunk • Limbs • Twigs • Buds • Fruits • Leaves • Insects • Bird nest • Bird eggs • Birds • Rain drops • Soil • Grass • Decay	• Roots anchored to soil • The soil holds the grass • Root connected to trunk • Trunk connected to limb • Leaves hang off twigs • Insects are on twigs • Bird nest is on a limb • Eggs are in the bird nest • Rain drops are on some leaves, twigs, grass, and in the soil • Some leaves lay on the grass • Embryos in eggs • Insect on a leaf	• Roots absorb water & minerals • Trunks distribute product from roots & leaves • Limb supports the bird nest & eggs • The tree provides habitat for birds and insects • Birds are perched on a twig • The egg embryos in the nest are developing into young birds • Tree leaves are producing sugar through photosynthesis • Carbon is released from the decaying leaves on the grass • On a leaf an insect is drinking

Classification of organisms and trophic levels in ecosystems (see Figure 3-8)

Energy flows and matter recycling (see Figure 3-9)

Aerobic Respiration: The net chemical change for aerobic respiration is the opposite of that for photosynthesis

Aerobic Respiration p.64	Photosynthesis p.73

Tolerance Ranges of Species

Law of tolerance	Acclimation	The threshold effect

Limiting factors in ecosystems

Roles (function) of Species in Ecosystems

Types of species in ecosystems

Type of Species	Role (function)
Endemic	
Immigrant or Exotic	
Indicator species	
Keystone species	

Ecological niche

Interactions between species

Energy Flow in Ecosystems

Food chains and food webs

Food chain (see Photo 3-12)	Food web (see Figure 3-12)

Trophic levels (see Figure 3-8)

Productivity of producers

Biomass	Gross primary productivity	Net primary productivity

Matter Cycling in Ecosystems

Nutrient cycles: *If an organism is to live, grow, and reproduce, it must take in variable amounts of different nutrients.*

Macronutrients	Micronutrients

Nutrient cycles or biogeochemical cycles are multidirectional

Carbon and Oxygen Cycle *(photosynthesis and respiration, see Figure 3-13)*

Nitrogen cycle (see Figures 3-13 and 3-14)

Nitrogen-Fixation	
Nitrification	
Assimilation	
Ammonification	
Denitrification	

Phosphorus Cycle (see Figure 3-15)

Hydrological Cycle (see Figure 3-16)

Rock Cycle: (Note: Limestone deposits are found in carbonate bedrock, pH>7)

Igneous rock	Sedimentary rock	Metamorphic rock

Terrestrial and Aquatic Ecosystems

The geography of life: *For centuries, people have been fascinated by geographic variations in the kinds and numbers of species found in various parts of the world.*

Life on land: Major terrestrial biomes

Biome	Characterization			
	rainfall	temperature	plant	animal
Permanent Ice				
Tundra				
Taïga or boreal forest				
Temperate deciduous forest				
Temperate rain forest				
Temperate shrub land (chaparrals)				
Temperate grasslands				
Tropical rainforests				
Tropical seasonal forests				
Deserts				

Climate and vegetation vary with latitude & altitude (i.e. vertical stratification of trees)

Life on Earth: Major aquatic biomes

Oceans (see also Figure 3-18)

Feature	Ocean Ecosystems
Coastal zone	
Coral reefs	
Estuaries	
Coastal wetlands	
Mangrove swamps	
Barrier Islands	
Benthic zone	
Pelagic zone	

Freshwater ecosystems

Freshwater lake types (see also Figure 3-19)

Eutrophic	Mesotrophic	Oligotrophic

Freshwater Rivers and Streams

Inland wetlands

Response to Environmental Stress

The constancy of change: *The "balance of nature" is a misnomer – the key thing happening in nature is change, constant change, caused by both natural and human-related forces and adjustments to environmental stresses.*

Changes in population size

Biotic Potential	Environmental resistance

Carrying capacity

Changes in population size: "J" and "S" curves (Figure 3-20)

"J" curve	"S" curve

Biological evolution, adaptation, and natural selection

Speciation and extinction

Ecological succession

Primary succession	Secondary succession

Human impacts on ecosystems
Principle of connectedness

Working with nature

TRACKING YOUR PROGRESS
Chapter 3—Earth's Life Support Systems

Multiple Choice Questions

1. What is Earth's matter?
   a) elements and compounds
   b) closed systems and open systems for matter
   c) mass and space
   d) electrons and protons
   e) salts, sugars, acids, bases and rocks

(p. 51)

2. Atoms consist of subatomic, electrically charged particles called:
   a) atoms only
   b) neutrons only
   c) mass
   d) electrons only
   e) ions

(p. 51)

3. The following represents one molecule of glucose:
   a) NH_3
   b) $C_6H_{12}O_6$
   c) $NaCl$
   d) CH_4
   e) $C_{12}H_{12}O_6$

(p. 53)

4. The following contains only one atom of oxygen:
   a) H_2O
   b) NO
   c) $2CO$
   d) (b) and (c) only
   e) all of the above

(p. 53)

5. The "mass number" of uranium 238 (isotope):
    a) 146n
    b)   92p
    c)   92e
    d) 238
    e)   92
(Pp. 52-53)

6. An example of potential energy is:
    a) moving trains
    b) heat
    c) hamburger
    d) wind
    e) streams
(p. 54)

7. The following is an example of low-quality energy:
    a) electricity
    b) Saudi Arabia's oil deposits
    c) propane
    d) uranium
    e) a hamburger
(p. 54)

8. During a physical or chemical change energy is neither created nor destroyed:
    a) First law of biodiversity
    b) Second law of kinetic energy
    c) Third law of conservation of matter
    d) Second law of thermodynamics
    e) First law of thermodynamics
(p. 54)

9. Oxygen accounts for _____ % of the volume of dry air.
    a) 21
    b) 34
    c) 66
    d) 78
    e) 99.9
(Pp. 55, 57)

10. If the ozone layer were a pile of dimes, it would be as thick as _____ Canadian dimes.
    a) 10 new 2001
    b) no
    c) 20 pre-1960
    d) 3
    e) more Canadian dimes than one can ever count in a lifetime
(p. 55)

11. The term _____comes from the Greek words *oikos* and *logos* (study of):
    a) entomology
    b) ethnology
    c) biology
    d) geomorphology
    e) ecology
(p. 57)

12. A group of individuals of the same species living & interacting in the same area at the same time is called a(n) _____.
    a) population
    b) habitat
    c) organism
    d) species diversity
    e) genetic diversity
(p. 58)

13. Which of the following is a prokaryotic organism?
    a) human
    b) bacteria
    c) protozoa
    d) slime mold
    e) protista
(p. 61)

14. Organisms that produce energy through the process of aerobic respiration are:
    a) consumers
    b) producers
    c) chemotrophs
    d) both (a) and (b) are correct
    e) all of the above
(Pp. 62-64)

15. Feeders who ingest or engulf particles, parts, or whole bodies of others are called:
    a) microconsumers
    b) macroconsumers
    c) chemotrophs
    d) autotrophs
    e) producers
(p. 63)

16. In the ecological sense, the eating habits of pigs are like:
    a) fungi – decomposers
    b) voles – herbivores
    c) deer – herbivores
    d) wolves – carnivores
    e) humans – omnivores
(Pp. 63-64)

17. An example of a primary consumer is a(n) _____.
    a) alligator
    b) wolf
    c) giraffe
    d) eagle
    e) rose
(p. 63)

18. A pregnant boa constrictor, which ate a mouse, would be considered:
    a) a primary consumer
    b) a secondary consumer
    c) a tertiary consumer
    d) a producer
    e) none of the above
(p. 63)

19. Maggots, dung beetles, shrimp, earthworms, and humans are all _____.
    a) carnivores
    b) decomposers
    c) detritivores
    d) scavengers
    e) dirt feeders
(p. 63)

20. Decomposers are generally considered to be:
    a) fungi
    b) bacteria
    c) plants
    d) (a) and (b) only
    e) all of the above
(p. 63)

21. As energy moves up the trophic levels, from plants to consumers, its availability:
    a) increases
    b) decreases
    c) remains stable
    d) there is no energy here
    e) doubles at each levels as more is created through the process
(Pp. 63-64)

22. The main ecosystem structure linked by energy flow and matter recycling is _____.
    a) sun energy
    b) energy, chemicals, and organisms
    c) sun energy and chemicals
    d) chemicals and organisms
    e) photosynthetic products
(p. 64)

23. This equation illustrates the typical process involved in photosynthesis:
    a) $C_6H_{12}O_6 + 6 O_2 = 6H_2O + 6CO_2 + energy$
    b) $H_2O + sunlight + CO_2 = energy + O_2$
    c) $glucose + O_2 = CO_2 + H_2O$
    d) $H_2O + energy = C_6H_{12}O_6 + sunlight$
    e) $6H_2O + 6CO_2 + solar energy = C_6H_{12}O_6 + 6 O_2$
(p. 64)

24. The net chemical change in aerobic respiration is the opposite of that for:
    a) the destruction of energy
    b) anaerobic respiration
    c) decomposition
    d) protein-synthesis
    e) photosynthesis
(p. 64)

25. If rainbow trout die suddenly in droves, under thick ice, this would be an example of:
    a) acclimation
    b) threshold effect
    c) limiting factor principle
    d) (b) and (c) are both correct
    e) all of the above
(p. 65)

26. Zebra mussels have devastating effects on the Great Lake crayfish. Zebra mussels are:
    a) an exotic species
    b) endemic to the Great Lakes
    c) an indicator species of the crayfish
    d) a keystone species of the Great Lakes
    e) both (a) and (d) are correct
(p. 66)

27. A young wolf chasing a butterfly near its den is an example of:
    a) an ecological niche
    b) chemical potential
    c) law of tolerance
    d) threshold effect
    e) two specialized species
(p. 66)

28. Organisms such as Pandas and Koalas are restricted to certain food types are:
    a) indicator species
    b) specialist species
    c) generalist species
    d) keystone species
    e) bellwether species
(p. 67)

29. A competing species that must move away or die out is an example of:
    a) fundamental niche
    b) interspecific competition
    c) competitive exclusive principle
    d) resource partitioning
    e) realized niche
(p. 68)

30. Canadian Prairie gophers are:
    a) exotic
    b) keystone species
    c) predators
    d) pets
    e) creators

(Pp. 66-67)

31. An epiphyte growing on a very large old-growth Douglas fir is a:
    a) predator of large trees
    b) species with a large appetite
    c) mutualism, highly social plant
    d) commensalism, a symbiotic relationship
    e) potential cancer treatment

(p. 69)

32. The Pacific yew tree, once a trash tree and now a new source for treating ovarian cancer, is a:
    a) specialist species
    b) limiting factor
    c) symbiotic carnivore
    d) generalist species
    e) nitrogen fixing lichen

(Pp. 68-69)

33. These animals exhibit resource partitioning by exhibiting stratified feeding behaviours.
    a) deer and grasshoppers
    b) blue jays and bald eagles
    c) cougars and mountain goats
    d) Richardson's ground squirrels and burrowing owls
    e) various species of *Dendroica* species (wood warblers)

(p. 69)

34. The second trophic level in "algae, mayfly nymph, salmon, grizzly bear, and you" is:
    a) you
    b) mayfly nymph
    c) grizzly bear
    d) salmon
    e) algae

(Pp. 64, 70)

35. The path of energy flowing through a boreal forest is:
    a) circular
    b) cumulatively increased
    c) exhausted by the third trophic level
    d) linear
    e) the meeting of two links
(p. 70)

36. Dead animals and plants (detritus) are NOT part of:
    a) food webs
    b) ecosystems
    c) ecological niches
    d) all of the above
    e) none of the above
(Pp. 70-71)

37. A snake swallowing a Mallard duck egg in 22 seconds is probably:
    a) a fast meal
    b) total biomass
    c) gross primary productivity
    d) net primary productivity
    e) respiration
 (p. 70)

38. Examples of micronutrients are:
    a) nitrogen, carbon, oxygen
    b) hydrogen, oxygen, molybdenum
    c) zinc, boron, molybdenum
    d) zinc, iron, nitrogen
    e) calcium, iron, phosphorus
(p. 71)

39. The following is NOT part of the nutrient cycles:
    a) toxins
    b) solar energy
    c) gravity
    d) all of the above
    e) none of the above
(p. 72)

40. Ocean kelp removing CO_2 from water represents which cycle?
    a) nitrogen cycle
    b) water cycle
    c) phosphorus cycle
    d) carbon cycle
    e) rock cycle
(Pp. 73-74)

41. Denitrification converts:
    a) ammonia (NH_3) to atmospheric nitrogen (N_2)
    b) atmospheric nitrogen (N_2) to ammonia (NH_3)
    c) ammonia (NH_3) to nitrates (NO_3)
    d) nitrates (NO_3) to atmospheric nitrogen (N_2)
    e) nitrous oxide to a super car engine fuel
(Pp. 73-74)

42. The following never occurs in the atmosphere:
    a) phosphorus
    b) carbon
    c) hydrogen
    d) oxygen
    e) nitrogen
(p. 74)

43. The following increase seepage (water recharge in groundwater supplies):
    a) clear-cutting trees
    b) healthy mature trees
    c) run-off
    d) erosion
    e) landslides
(p. 75)

44. If global warming increases the rainfall & summer temperatures of the taiga, we get a:
    a) tropical rainforest
    b) boreal forest
    c) temperate woodland
    d) temperate forest
    e) tundra
(p. 77)

45. Here you may find zooplankton, baleen whales, herring, and sardines:
    a) profundal zone
    b) littoral zone
    c) euphotic region
    d) prokaryotic zone
    e) benthic environments
(Pp. 80-81)

46. An inland lake with lots of Zebra mussels, low levels of nitrates and phosphates is:
    a) eutrophic
    b) oligotrophic
    c) euphotic
    d) autotrophic
    e) detritotrophic
(p. 82)

47. This ecosystem has cold water with high oxygen levels and clinging organisms:
    a) lake
    b) river
    c) stream
    d) wetland
    e) pond
(p. 82)

48. The maximum rate (rmax) at which a population could increase is:
    a) carrying capacity
    b) limit to growth
    c) biotic potential
    d) "S" curve
    e) environmental resistance
(p. 83)

49. What experiment caused some scientists to pause and think about how ecosystems work?
    a) Biosphere II
    b) Lithosphere II
    c) Ecosphere II
    d) Hydrosphere II
    e) Atmosphere II
(Pp. 84-85)

50. Change in inherited characteristics of a population from generation to generation is:
    a) emigration and immigration – population dynamics
    b) artificial cloning (plants/animals) by research labs - biological experimentation
    c) routine selective breeding of domestic stock for specific traits – agriculture
    d) phenotypic variation among organisms – environmental sciences
    e) driving force of adaptation to environmental change - biological evolution
(p. 85)

51. Excessive radiation (X-rays, ultraviolet light) and certain toxic chemicals may cause:
    a) our skin to toughen
    b) make us very huge
    c) mutation of our DNA
    d) artificial selection to be unethical
    e) DNA genes pools in our cells
(p. 85)

52. Mass extinction killed off:
    a) auks
    b) dinosaurs
    c) passenger pigeons
    d) milk snakes
    e) black-tailed deer
(p. 86)

53. A White pine old growth forest:
    a) climaxed community
    b) primary succession
    c) secondary succession
    d) dead forest
    e) forest with the most biodiversity ever!
(p. 87)

54. To achieve sustainability we must first:
    a) manage ourselves
    b) manage resources
    c) learn to live apart from nature – in cities
    d) (a) and (b) are both correct
    e) all of the above
(p. 87)

CHAPTER 4—HUMAN POPULATION ISSUES AND THE ENVIRONMENT
(pp. 91-115)

STUDENT LEARNING OBJECTIVES

After studying this chapter you should be able to:
- explain why some people consider the human population issue to be the most important environmental issue
- understand how population size is affected by rates of birth, death, fertility, and migration
- discuss the importance of exponential growth rates, doubling times, and other measures of population dynamics
- describe the four major phases in the demographic transition
- construct an age structure diagram and use it to predict future population growth rates
- discuss the approaches taken in response to world population growth
- define and discuss all of the KEY TERMS listed below

KEY TERMS (defined at the page number shown and in the glossary)

NOTES

Introduction
Pollution of natural environments, declining energy reserves, reduced biodiversity, and wildlife extinctions are related to human population growth.

Basic Population Concepts

Population and technology

Developing and developed countries (see Box 4-1)

Developing countries	Developed countries

Human demography *(see <u>Boom Bust and Echo</u>, by Foot & Stoffman, 1996, for more ideas)*

Human population growth *(see Brief History of the World's Growth Stages, Table 4-1 and <u>Ishmael</u> by Daniel Quinn, 1992 for more ideas):*

Stage	Time Period	Population Density	Total Human Population	Average Rate of Growth
1. Hunters and Gratherers (Leavers*)				
2. Early pre-industrial agriculture				
3. The machine age (Industrial Revolution)				
4. The modern era				

Population dynamics

CGR = CBR – CDR

C =

G =

R =

B =

D =

Explain the above equation:

World population milestones and vital events (see Table 4-2, contrast with Table 4-3)

Predicted world population	Year	Time span
1 billion	1804	--
2 billion		123 years later
3 billion		
4 billion	1974	
4 billion		
5 billion		
6 billion	1999	12 years later
7 billion		
8 billion	2028	14 years later

Exponential growth *(see also Figure 3-20, Table 4-3)*

Projecting Future Population Growth

Doubling time (T) – *(see Figure 4-2)*

T = 70/annual growth rate (%)

China	T	=	70/_____	(%)	=	_____
India	T	=	70/_____	(%)	=	_____
Canada	T	=	70/_____	(%)	=	_____
France	T	=	70/_____	(%)	=	_____
West Bank	T	=	70/_____	(%)	=	_____

Notes:

Logistic growth curve

Logistic growth curve: *"S"and "J" curves (see figure 4-3)*

Demographic Transition
The four-stage model (see Figure 4-4)

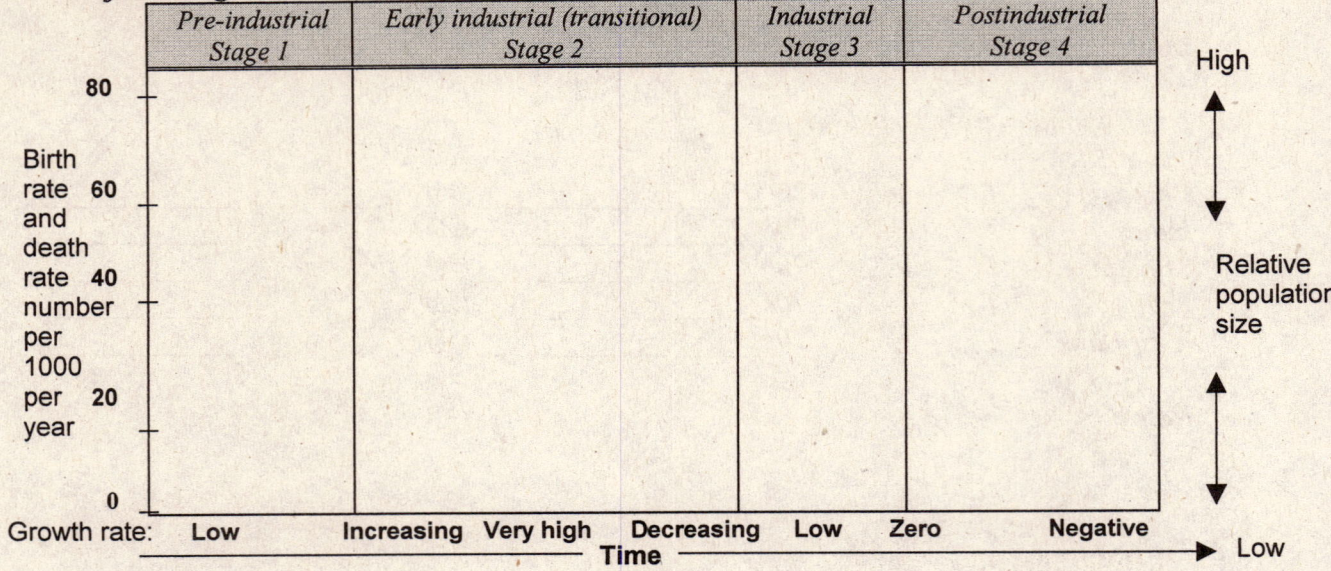

Demographic trap (p. 100)

Diseases and death in industrial society
Modern medicine

Zero population growth

Carrying capacity

Limiting factors - *Per capita availability of resources and Thomas Malthus (1798)*

Age structure - *Population age structure diagrams Figure 4-5 and Figure 4-6*

The dependency ratio

Fertility rates and lag-time effects (*see Table 4-5, Figure 4-7*)

General fertility rate	Specific fertility rate	Total fertility rate (TFR)	Replacement fertility

Future population trends

Cultural factors

Life expectancy

Facing the Problems of World Population Growth

Increasing the education levels and the marriage age

Marriage age and number of children born in Sri Lanka vs. Bangladesh

Sri Lanka		Bangladesh	
Marriage age	Number of children	Marriage age	Number of children

Birth control in developing nations – *family planning education helps women to control family size and the interval between births.*

National birth rate reduction programs
India:

China:

Migration

Population and environmental sustainability

TRACKING YOUR PROGRESS
Chapter 4—Human Population Issues and the Environment

Multiple Choice Questions

1. The world population was _____ billion in 1980, and _____ billion in 1999.
   a) 1, 10
   b) 2, 5
   c) 5, 20
   d) 4, 6
   e) no one really knows the approximate number of humans on Earth

(p. 91)

2. Which of the following is NOT a characteristic of a LDC?
   a) birth rates are higher than the level needed to sustain the population
   b) per capita incomes are high
   c) the level of industrialization is low
   d) infant mortality rates are high
   e) short life expectancies

(p. 92)

3. Thailand could be classified as an:
   a) LDC
   b) MDC
   c) HDC
   d) GDC
   e) ABC

(p. 92)

4. Paul Ehrlich is a well-known American expert _____:
   a) oceanographer
   b) geographer
   c) population biologist
   d) ecologist
   e) both (c) and (d) are correct

(p. 92)

5. A timed video of Ste. Catherines Street, Montreal circa 1901, 1952, 2000, would show:
    a) a country side road, a few homes, then a nice "green" street
    b) a nice "green" street, replaced by street-cars, then a crowed congested traffic
    c) busy street-cars, replaced by a nice "green" street, turned into a crowded street
    d) some homes, then a nice "green street", replaced by a country side road
    e) a nice "green" street, replaced by street cars, then a crowed, congested street
(Pp. 91-92)

6. We use this to improve our understanding of human population impacts on the biosphere:
    a) actuarial science
    b) cartography
    c) numerology
    d) demography
    e) geographic information systems
(p. 93)

7. The collection of personal information about age, gender, and livelihood is:
    a) a census
    b) no one's business
    c) for estimating birth and death rates
    d) only sought out in industrialized society
    e) only sought out in developing nations
(p. 93)

8. Which stage included the domestication of animals and the rise of settled villages:
    a) one, the hunter-gatherer society
    b) two, early, preindustrial agriculture
    c) three, the industrial revolution
    d) four, the modern era
    e) none of the above
(p. 94)

9. What is the predicted world population in 2028?
    a) 1 billion
    b) 6 billion
    c) 8 billion
    d) 20 billion
    e) 120 billion
(p. 95)

10. If CDR is 4.01% and the CBR is 5.02%, then the CGR of the population is:
    a) 9.03%
    b) 1.01%
    c) 20.02%
    d) -1.01%
    e) 1.25
(p. 95)

11. The infant mortality rate is:
    a) the annual number of live births per 1000 population
    b) the annual number of dead babies per 1000 population
    c) the annual number of abortions per 1000 population
    d) the ratio of deaths of infants under 12 months of age per 1000 live births
    e) the ration of abortions and live births under 12 months of age per 1000 mothers
(p. 95)

12. It is expected that developed regions will experience a _____ growth rate and their absolute numbers will _____ by about 2040.
    a) positive, increase
    b) negative, increase
    c) negative, decrease
    d) positive, decrease
    e) negative, stabilize
(Pp. 95-96)

13. During the first half of the 20th century, population actually increased at a:
    a) faster rate than an exponential rate
    b) slower rate than an exponential rate
    c) at an exponential rate
    d) at an arithmetical rate
    e) at a mathematical rate
(p. 96)

14. Doubling time can be estimated by the formula:
    a) T = 70/annual growth rate (%)
    b) T = 50% + birth rate + death rate
    c) T = CBR – CDR (%)
    d) T = 70 + birth rate + death rate (%)
    e) T = 50 x 70/annual birth – deaths (%)
(p. 97)

15. What is the population doubling time if the annual population growth equals 3?
    a) 100 years
    b) 124 years
    c) 23 years
    d) 2 years
    e) this can NOT be calculated
(p. 97)

16. *The Limits to Growth,* by D.H. Meadows et al. (1972), was influential in challenging:
    a) society about food availability
    b) society's economic decline during the war
    c) society's need for 5+ children per family
    d) society about the availability of solid education and medicine
    e) society's thinking about planetary limits
(p. 98)

17. The human logistic growth curve is:
    a) "S" shaped
    b) "J" shaped
    c) "M" shaped
    d) "W" shaped
    e) "Z" shaped
(p. 99)

18. The effect of this feature greatly influences Canada's growth rate & doubling time:
    a) birth rate
    b) death rate
    c) immigration and emigration rates
    d) life expectancy
    e) refugee claims
(p. 99)

19. All highly developed nations with advanced economies have gone through a

_____.
    a) demographic transition
    b) demonstrative trap transition
    c) demoralization transition
    d) decongestion trap
    e) death transition- trap
(Pp. 99-100)

20. The preindustrial stage is characterized by:
    a) good living conditions, low birth rates, high death rates
    b) good living conditions, high birth rates, low death rates
    c) harsh living conditions, low birth rates, high death rates
    d) neutral living conditions, zero population growth
    e) harsh living conditions, high birth rates, high death rates
(p. 100)

21. About 90% of children now infected with HIV were born in:
    a) India
    b) Asia
    c) Japan
    d) Africa
    e) Bangladesh
(p. 101)

22. When we determine, and agree, on an average standard of living and quality of life:
    a) all of humanity will suffer
    b) the carrying capacity for humanity will be calculated
    c) human impacts on the environment will take their toll
    d) all of the above
    e) none of the above
(p. 102)

23. Who said: "The power of population is greater than the power in Earth to produce food for man"?
    a) Savard 10 AD
    b) Plato 250 BC
    c) Mathus 1798
    d) Quinn 1996
    e) Draper 2001
(Pp. 102-103)

24. Off the coast of North Carolina, toxic _____ appear due to _____.
    a) algal blooms, nutrient rich fertilizers flowing to the sea from farms
    b) dinoflagellates, nutrient-rich human and animal wasted flowing into the sea
    c) bacteria, decomposing human waste
    d) multicellular parasites, immeasurable numbers of dead sea animals and fish
    e) both (a) and (d) are correct
(Pp. 102-104)

25. Launching a successful toy store versus a greenhouse store may be predicted from:
    a) age structure diagrams
    b) population indices
    c) pre-productive cycle-grams
    d) post-reproductive histograms
    e) pre-reproductive graph interpretation
(Pp. 103-109)

26. Current responses to ageing populations in developed countries include:
    a) a larger proportion in the younger age classes
    b) child care facilities for grandchildren
    c) gated retirement communities
    d) a smaller proportion in both the oldest and youngest age classes
    e) population momentum
(p. 108)

27. What event occurs for many years after a population achieves replacement fertility?
    a) population lag effect
    b) limiting factor
    c) zero population
    d) population momentum
    e) (a) and (d) are both correct
(p. 108)

28. The TFR for Afghanistan is 6.1, and 3.0 for Mexico. What is the TFR for Canada?
    a) 2.0
    b) 10.0
    c) −2.0
    d) 1.5
    e) unknown
(p. 108)

29. Compared with stable populations, increasing populations have:
    a) a smaller proportion in the younger age classes
    b) an equal proportion in all age classes
    c) a smaller proportion in both the oldest and youngest age classes
    d) the widest age classes half way up the pyramid
    e) a smaller proportion in the older age classes
(Pp. 108-109)

30. Life expectancy for the human population has _____ since ancient times.
    a) NOT increased
    b) increased by 20 years
    c) doubled
    d) actually decreased by 20 years
    e) decreased two fold
(p. 110)

31. What has exacerbated issues relating to food supply, land and soil resources, and water?
    a) forest product demands
    b) global warming
    c) growth numbers of the human species
    d) huge caribou herds in the Arctic
    e) acid rain
(p. 110)

32. Typically, women with more education tend to:
    a) marry earlier and have more children
    b) marry later and have fewer children
    c) marry later and have more children
    d) NOT marry and have more children
    e) NOT marry and have fewer children
(p. 111)

33. Since the "one child per family law", _____ neonate females went missing in China.
    a) three to four hundred
    b) three to four thousand
    c) 30 to 40 thousand
    d) 30 to 40 million
    e) 30 to 40 billion
(Pp. 112-113)

34. In some countries, this has surpassed employment as a leading reason for immigration:
    a) political rights
    b) refugee status
    c) freedom of expressions
    d) natural spaces
    e) family reunification
(p. 113)

CHAPTER 5—OUR CHANGING ATMOSPHERE
(pp.118-165)

STUDENT LEARNING OBJECTIVES

After studying this chapter you should be able to:
- understand the main issues and concerns relating to Canada's atmosphere
- identify a range of human uses of the atmosphere
- describe the impacts of human activities on the atmosphere around us .
- appreciate the complexity and interrelatedness of issues relating to the atmosphere
- outline Canadian and international responses to the need to protect the atmosphere
- discuss challenges to a sustainable future for the atmosphere
- define and discuss all of the KEY TERMS listed below

KEY TERMS (defined at the page number shown and in the glossary)

biomass burning	124	phenology	142
Dobson units	132	polar vortex	136
global warming potential (GWP)	127	polar stratospheric clouds	136
halocarbons	123	precursor chemicals	133
halons	125	volatile organic compounds	149

NOTES

Introduction
"We, the members of the National Forum on Climate Change, believe that climate change will touch the life of every Canadian."

Recent causes of climate change *(see Table 5-1 and Table 5-2)*

Earth's natural climate system

The global climate system
Natural greenhouse effect

Changes in Atmospheric Composition

Enhanced greenhouse effect

Atmospheric gas	Changes
Carbon dioxide	
Methane	
Nitrous oxide	
Chlorofluorocarbons and Halons	
Aerosols*	

*Aerosols from aircraft emissions may be becoming more important in both the troposphere and stratosphere

Radiative and residence characteristics of greenhouse gases

Effects of enhanced greenhouse gases

Selected impacts of enhanced greenhouse gases (see Table 5-4)

Area of Change	Description or comment
Air temperature	
Sea-level rise	
Natural ecosystems	
Forests	
Agriculture	
Economic activities	
Human health	

Thinning of ozone layer
The ozone layer (see Figure 5-3)

Changes in the ozone layer

Anthropogenic Ozone-Depleting Substances (ODS)

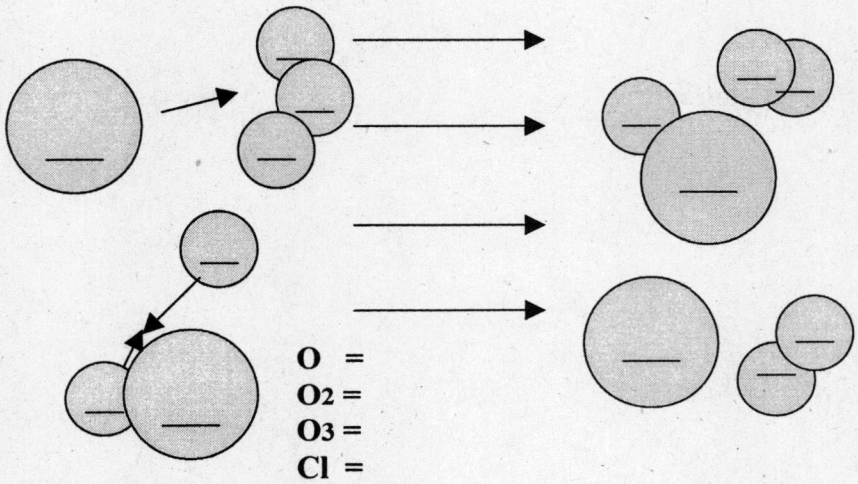

O =
O2 =
O3 =
Cl =

How ozone depleting substances destroy stratospheric ozone (Figure 5-4)

Natural ozone-depleting substances

Spatial Variation in Ozone Depletion

Antarctic ozone depletion: There is NO hole in the ozone shield around the Earth, but there is a decrease in the concentration of ozone that occurs during the Antarctic spring.

Tropical and mid-latitude ozone depletion

Effects of ozone depletion

Increased ultraviolet radiation

The UV index (see "slip, slap, slop", Table 5-5 and Enviro-Focus 5)

Other Atmospheric Changes

ENSO and PDO and other major surface ocean currents affecting climate

Surface current	Selected characteristics (see Table 5-6)
Arctic Circumpolar Wave	
Pacific Decadal Oscillation (PDO)	
El Niño & the Southern Oscillation (ENSO)	
North Atlantic Oscillation	
Tropical Atlantic Variability	

Acidic deposition

What is acid rain? (see Figure 5-7)

Effects of acidic deposition

Aspect Affected	Nature of Effects (see Table 5-7)
Aquatic ecosystems	
Terrestrial plant life	
Animal life	
Socioeconomic consequences	
Human health	

Airborne contaminants
Canadian high Arctic

Cities and air quality

Predicting climate change
Computer models (see Table 5-8)

Responses to Atmospheric Changes

International actions

Controlling greenhouse gas emissions

The Kyoto Protocol on Climate Change (see Table 5-9)

Protecting the ozone layer

Controlling acidic deposition

The Canadian Council of Ministers of the Environment (see Box 5-6)

Canadian Law, Policy and Practice

CEPA (1999)

NAPCC (1995)

Canadian Partnerships and Local Actions

Future Challenges

TRACKING YOUR PROGRESS
Chapter 5—Our Changing Atmosphere

Multiple Choice Questions

1. Changes in the composition of our atmosphere began when fossil fuels ignited the:
   a) Agricultural Revolution, 12 000 years ago
   b) Industrial Revolution, 1750s
   c) Second World War, 1940s
   d) Cold War, 1960s
   e) Desert Storm, 1990s
(p. 119)

2. Burning of large quantities of fossil fuel can introduce:
   a) CO_2
   b) CO_2 and CO
   c) CO_2, CO, and SO_2
   d) CO_2, CO, SO_2, and NO_x
   e) CO_2, CO, SO_2, NO_x, CH_4, and more
(p. 120)

3. What event occurs when we change the amount of atmospheric solar input and ozone?
   a) climate forcings
   b) atmospheric cataclysms
   c) stratospheric rebounds
   d) Earth paradigm shifts
   e) no event occurs, as we can not cause such changes
(Pp. 120-121)

4. What study suggested low concentrations of greenhouse gases during glacial periods?
   a) Geological Surveys of Canada, 1904
   b) International Society of Climatologists, 1997
   c) Intergovernmental Panel on Climate Change, 1995
   d) Canadian Geographical Society, 2000
   e) Drilling and blasting records at Frost, 1967-2001
(p. 122)

5. Measurements of CO2 are taken at Mauna Loa Observatory, HI, _____ urban areas.
  a) within various
  b) next to
  c) under some
  d) above major
  e) far from
 (Pp. 123-124)

6. Among the largest anthropogenic sources of methane are cultivated:
  a) crawfish and rice
  b) tulapia and escargots
  c) rice paddies and livestock
  d) hemp and corn
  e) salmon and herring
(p. 124)

7. Between 1980 and 1995 global warming agent sulphur dioxide emissions have:
  a) doubled
  b) tripled
  c) remained the same
  d) decreased
  e) no one knows what they've done
(p. 125)

8. What is the main source of black carbon aerosols in tropical and southern
   hemispheres?
  a) hair sprays and lubricants
  b) exhaust from lead gas emissions
  c) biomass burning
  d) tropical rainforest fires
  e) smoke signals for world peace
(p 126)

9. This index takes into account the different times that gases remain in the atmosphere.
  a) HCFC
  b) GWP
  c) CH
  d) Halons-HCFCs
  e) radiative characteristic analysis by computer modelling techniques
(p. 127)

10. CO_2 is the least effective greenhouse gas per kilogram emitted, but _____.
    a) its contribution to global warming is the least
    b) its contribution to global warming is the largest
    c) its contribution to ozone depletion is the least
    d) its contribution to ozone depletion is the largest
    e) its contribution to salmon depletion is huge
(p. 127)

11. Every year the world grows by this many humans:
    a) 1.2 thousand
    b) 6 million
    c) 90 million
    d) 2.3 billion
    e) 3.4 trillion
(Pp. 127-128)

12. Among the uncertainties regarding greenhouse gases are concerns about:
    a) oceans & terrestrial ecosystems and their role in absorption of CO_2 emissions
    b) oceans & terrestrial ecosystems' ability to produce new carbon sinks
    c) oceans & terrestrial ecosystems and their role in absorption of NO_2 emissions
    d) oceans & terrestrial ecosystems' depleting carbon monoxide sinks
    e) (a) and (c) are both correct
(Pp. 128-129)

13. The polar ice cap has thinned by _____ per cent in three decades.
    a) 4
    b) 10
    c) 40
    d) 65
    e) 99
(Pp. 128-131)

14. The symbol O_3 represents:
    a) oxygen gas
    b) oxide
    c) oxygen triplets
    d) laughing gas
    e) ozone
(Pp. 132, 134)

15. This is equivalent to three hundred (300) parts per billion:
    a) peak ozone concentrations between 20 and 25 km above the Earth
    b) peak Dobson units over the Arctic
    c) minimal ozone concentration required to prevent extinction of humans
    d) maximal ozone output achievable on Earth
    e) atmospheric ozone danger
(p. 132)

16. The amount of stratospheric ozone depletion attributed to CFCs alone is:
    a) 20%
    b) 50%
    c) 65%
    d) 80%
    e) 99%
(p. 133)

17. The following synthetic compounds are all involved in ozone depletion:
    a) butane, fizz from soda pops like Coke and Pepsi, FCFs
    b) CO_2, CH_4, and CCF
    c) uranium, basalt and calcium chloride gases
    d) propane, butane, and gasoline
    e) halons, methyl chloroform, methyl bromide, and carbon tetrachloride
(p. 133)

18. HCFCs are most likely to be found in:
    a) paper products, especially coated cardboard and newspapers
    b) refrigerants, degreasers, foams and sterilant gas mixtures
    c) hydro transformers, fluorescent light ballast, and batteries
    d) plastic wraps, bags, and other plastic packaging
    e) film containers, car mats and insulation
(p. 133)

19. Chlorine acts to eliminate ozone by:
    a) causing ozone molecules to precipitate out of the stratosphere
    b) firing a chemical reaction which shoots ozone molecules into space
    c) combining with O_3 to cause heavy ground-level ozone
    d) producing chlorine monoxide and oxygen from ozone
    e) a chemical process NOT yet understood
(p. 134)

20. The hole in the ozone is:
    a) visible and increasing slowly
    b) visible and shrinking
    c) visible and increasing fast (measured by Zonal-Ozone-O-Meters – ZOOMs)
    d) visible and stable
    e) NOT there as there is no "hole" in the ozone
(Pp. 135-136)

21. Ozone depletion might also affect:
    a) global warming by reducing populations of marine plankton which absorb
        large amounts of CO_2
    b) weather patterns by indirectly influencing cloud formation over the oceans
    c) ground-level air pollutants, and thus ameliorating human health
    d) phytoplankton
    e) (a) and (d) are both correct
(p. 136)

22. In Canada, _____ have been tested for UV-B sensitivity and 45% are affected.
    a) men and women over 21
    b) trees
    c) frogs
    d) rice fields
    e) insects
(Pp. 137-138)

23. The study of periodic occurrences in nature and their relationship to climate is called:
    a) ecology
    b) phenology
    c) anthropology
    d) periodology
    e) mysteries
(p. 142)

24. PDO temperature changes are likely behind historical shifts in:
    a) volcanic activity
    b) salmon stocks
    c) tectonic plate
    d) speciation
    e) climate
(p. 143)

25. A pH value of 3 is _____ times more acidic than a pH value of 5.
    a) 0.2
    b) 2
    c) 10
    d) 200
    e) 100
(p. 144)

26. This event affects the waxy surfaces of leaves, and in turn lowers its disease resistance:
    a) ozone depletion
    b) acid deposition
    c) increase in CFC, HCFC, and halon concentrations
    d) extra UV-B
    e) global warming
(Pp. 145-146)

27. The best soils and bedrock to reduce acidity of atmospheric deposition in Canada are:
    a) carbonate bedrock and/or deep fine textured soils
    b) permanent ice and snow fields
    c) noncarbonate bedrock and coarse textured shallow soils
    d) peatland and granite bedrock
    e) noncarbonate bedrock and/or shallow to deep soils
(p. 145-147)

28. Ground-level ozone, sulphur dioxide, and nitrogen dioxide can effect:
    a) human health
    b) rubber
    c) paint
    d) plastic
    e) all of the above
(p. 149)

29. Earth Summit +5 action was insufficient, but the lessons and challenges are clear as the international community prepares for the review of Agenda 21 in the year _____.
    a) 2002
    b) 2005
    c) 2008
    d) 2022
    e) NOT to be reviewed. The world leaders refused to discuss the environment.
(p. 151)

30. The Montreal Protocol signed by 24 nations to phase-out ODS came into effect in:
    a) July, 1978
    b) May, 1981
    c) September, 1987
    d) January, 1989
    e) December, 1996
(Pp. 154-155)

31. New smelting technology and extraction of sulphuric acid sufficiently permitted a:
    a) reduction of acid rain in Canada to 1990 levels
    b) bumper crop of fish in fresh-water commercial enterprises
    c) return to blueberry picking for fine Canadian pies and jellies
    d) extensive and successful revegetation program in Sudbury, Ontario
    e) revitalization of more than 200 000 lakes in Ontario and Québec
(p. 156)

32. CCME which stands for _____ , has a _____ Award program.
    a) Canadian Committee for Monitoring the Environment; Restoration
    b) Coastal Council for Mediating Efforts; Rehabilitation Incentive
    c) Countries Committed to Mother Earth; Permaculture
    d) Can Canada Mend the Environment; Green-up City
    e) Canadian Council of Ministers of the Environment; Pollution Prevention
(p. 157)

33. What act was developed in 1999 to provide Canada with tools for action on clean air?
    a) NGO
    b) CNA
    c) AQA
    d) CCME
    e) CEPA
(p. 158)

34. This product requires 30 to 40% less energy, creates 63 to 73% less air emissions, and 90% less water pollution than comparable paper bags.
    a) hemp magazine sleeves (ON's Dress-Lax program)
    b) plastic bags (BC's poly bags for magazine sleeves)
    c) nylon lunch bags (German lunch sacs)
    d) rice-paper sacs (Japanese mandatory trash containers)
    e) duvet (NWT's quality sleeping bags)
(p. 159)

CHAPTER 6—AGROECOSYSTEMS AND LAND RESOURCES
(pp. 167-204)

STUDENT LEARNING OBJECTIVES

After studying this chapter you should be able to:
- understand the main issues and concerns relating to Canada's land resources and their agricultural use
- identify a range of agricultural uses of our land resources
- describe the impacts of agricultural activities on land and affected water resources
- appreciate the complexity and interrelatedness of issues relating to land and agroecosystems
- outline some Canadian and international responses to the need for sustainable agriculture and agroecosystems
- discuss challenges to a sustainable future for land resources, agroecosystems, and agriculture
- define and discuss all of the KEY TERMS listed below

KEY TERMS (defined at the page number shown and in the glossary)

NOTES

Introduction
When agriculture is carried out in a sustainable manner, the natural resource base (land) is protected.

Canada's agricultural land base

Canada Land Inventory (CLI) (see Table 6-1)

Agroecosystems

Technological & socioeconomic changes in Canadian agriculture

Technological & socioeconomic changes in Canadian agriculture (continued...)

Monocultures	Summerfallow	Salinization

Use of Farmland in Canada (see Figure 6-3)

Use of farmland	Description of use	Area (ha millions)				
		1971	1976	1981	1986	1991
Improved pasture						
Summerfallow						
Cropland						
Improved cropland						
Total farmland						

Human Activities and Effects on Agricultural Lands

Effects on soil resources

Levels of soil organic matter

Wind, water, and tillage erosion

Soil structure

Soil salinization

Chemical contamination

Desertification

Effects on water resources

Contamination from crop production activities

Aquifers	Eutrophication

Contamination from livestock production activities

Walkerton, Ontario

Feedlots, manure management, and public health concerns (see Box 6-1)

Major issue	Problems
Feedlot operation	
Manure storage & application	
Pollution & human health problems	*Air quality:* *Surface water quality:* *Groundwater:*
Community stresses	
Actions to improve ILOs	

Irrigation effects

Effects on biodiversity
"...settlement and agricultural development have degraded the quantity and quality of wildlife habitat in Canada"

Wetland loss and conversion

Rangeland grazing

Riparian habitat disturbance

Cows and fish experiment (see Box 6-4)

Good stewardship of rangelands

Greenhouse gases

Item	Notes
Soils	
Methane	
Nitrous oxide	
Carbon dioxide	

Conservation tillage and zero tillage (see Box 6-5)

Energy Use

Responses to Environmental Impacts and Change
Globalization and Cash Crops

Regional sustainability

International initiatives

Canadian Efforts to Achieve Sustainable Agriculture

Examples of Manitoba's actions towards sustainable agriculture

Nontraditional agricultural activities

Organic farming	
Community shared agriculture	
Game farming & ranching	
Agricultural biotechnology (see Box 6-7)	

Partnerships

Future challenges

TRACKING YOUR PROGRESS
Chapter 6—Agroecosystems and Land Resources

Multiple Choice Questions

1. Canada's soil capability has less than 0.5% class 1 lands, while _____ per cent is
located in southern Ontario.
   a) 4
   b) 17
   c) 52
   d) 85
   e) 92
(p. 168)

2. Almost 60 per cent of Canadian land converted to urban use was formerly:
   a) prime agricultural land
   b) wetlands and marshes
   c) shallow silt on bedrock
   d) class 7 land
   e) mix of rocks and clay soil
(Pp. 169-170)

3. What is the cultivation of one crop over a large area known as?
   a) permaculture
   b) horticulture
   c) polyculture
   d) monoculture
   e) aboriculture
(p. 171)

4. Land NOT sown for at least 1 year to conserve moisture (enhances nitrogen content) is:
   a) improved cropland
   b) conservation tillage
   c) improved pasture
   d) cropland
   e) summerfallow
(Pp. 171-172)

5. Almost _____ per cent of Canada's agricultural production is exported:
 a) 50
 b) 20
 c) 10
 d) 0.5
 e) none
(Pp. 171-172)

6. About 36 per cent of cultivated land in the Prairies are subject to high to severe risk of:
 a) water erosion
 b) sudden prairie gopher infestation
 c) annual locust infestations
 d) wind erosion
 e) soil erosion
(Pp. 173-174)

7. Where is soil compaction NOT a serious problem?
 a) Quebec
 b) Ontario
 c) the West coast
 d) the East coast
 e) the Prairies
(p. 175)

8. Heavy metals from herbicides, insecticides, algaecides, and fungicides enter soils:
 a) from atmospheric deposition
 b) through fertilizers
 c) by way of animal manure
 d) from human sewage sludge and nutrients
 e) all of the above
(Pp. 175-176)

9. A combination of unsustainable land use practices and climatic variations resulted in:
 a) deforestation
 b) mining
 c) aquaculture
 d) desertification
 e) space exploration
(p. 176)

10. It's NOT good to allow cows to venture too close to waterways as manure bacteria may:
    a) enter the water-body by surface runoff
    b) be carried upwards by evaporation then fall down as precipitation
    c) be carried by cows into the water and let off by transpiration
    d) cause erosion
    e) be transmitted directly to producers by entering their oral cavities
(p. 176)

11. If phosphorus enters water bodies it can lead to accelerated:
    a) oligotrophication
    b) eutrophication
    c) malnutrition
    d) formation of rich, black humus
    e) Abbotsford aquifer formations
(p. 177)

12. These have more than 300 animal units confined in facilities at a density of 43 animal units per acre for more than 90 consecutive days, and manure must be managed.
    a) free ranging (FR) fields
    b) large paddocks (LP)
    c) intensive livestock operations (ILOs) on farms
    d) (a) and (b) are both correct
    e) none of the above
(p. 178)

13. This produces 9 954 kg of manure annually, including 56 kg of N, and 18 kg of P:
    a) 100 steers, 50 hogs, and 10 chicken
    b) 10 steers and 5 hops
    c) a single steer
    d) 5 hogs and 1 chicken
    e) a single chicken
(p. 179)

14. Algae blooms, accelerated weed growth, and eutrophication result from:
    a) heavy metals and carbon
    b) acid contamination and dung
    c) bio-magnification and surface runoff
    d) excessive phosphate and nitrogen
    e) contamination with iron and potassium
(p. 179)

15. In Alberta, the highest rates of intestinal infections are found in:
    a) ringbolts
    b) feedlot alleys
    c) people in the livestock industry
    d) (b) and (c) are both correct
    e) all of the above
(p. 180)

16. *Cryptosporidium parvum* may be spread by cattle and is resistant to:
    a) chlorination and cause human disease in Canada
    b) filtration and cause food crop disease in Canada
    c) fishers and is a fish disease found in Canada
    d) none of the above
    e) all of the above
(p. 180)

17. When was Alberta's Code of Practice for handling animal manures put into place?
    a) 1895
    b) 1925
    c) 1975
    d) 1995
    e) Will be in place by 2005.
(p. 181)

18. The Alberta Cows & Fish initiative is the ability of a riparian area to perform certain
    key ecological functions including:
    a) lining manure lagoons
    b) transportation of manure far away from ILOs
    c) testing soil for nitrogen and phosphorus
    d) monitoring collection lagoons to prevent overflow
    e) all of the above
(Pp. 181, 186-187)

19. Irrigation has traditionally been applied to fruit, tobacco, and vegetable crops in:
    a) Eastern Canada
    b) Central Canada
    c) Northern Canada
    d) Western Canada
    e) these items do NOT grow in Canada
(p. 181)

20. Alachlor, a carcinogenic and oncogenic pesticide banned in 1989, was found along
    the:
    a) Canadian shore of Lake Ontario in the late 1950s
    b) Canadian shore of Lake Huron in the late 1970s
    c) Canadian shore of Lake Superior in the early 1980s
    d) Canadian shore of Lake Erie in the early 1990s
    e) Canadian shore of Lake St. Clair in the early 2000s
(Pp. 183-184)

21. The major cause of riparian habitat disturbance is:
    a) roads and highways
    b) urban back-yards
    c) alley cats
    d) acid rain
    e) livestock grazing
(p. 186)

22. This is NOT a way that agricultural activities relate to greenhouse gas concentration:
    a) soil are an important natural source of, and reservoir for carbon
    b) methane is emitted from livestock and solid manure
    c) nitrous oxide is released from nitrogen fertilizers
    d) carbon dioxide is released from the burning of fossil fuels in framing activities
    e) these are ALL main ways that agriculture relate to greenhouse gas
        concentrations
(p. 188)

23. Prairie land farmers employ conservation tillage practices to fight against:
    a) soil erosion
    b) wind erosion
    c) water erosion
    d) (a) and (c) are both correct
    e) (b) and (c) are both correct
(Pp. 188-190)

24. The trend toward globalization indicates that food in our stores travels about ____ km.
    a) 20
    b) 2000
    c) 20 000
    d) 200 000
    e) NOT at all, it stays in the local community
(p. 191)

25. An alternative to globalization of foods is "regional sustainability". This means that developing countries would:
    a) grow crops for export only
    b) first grow food for the wold then crops for themselves
    c) first grow food for themselves then crops for export
    d) focus on experimental farming
    e) expand on holistic farming, never export crops
(p. 191)

26. This is a multinational and multidisciplinary project on women and resource management:
    a) Agenda 21
    b) WEDNET
    c) UNCED
    d) CIDA
    e) ELC
(Pp. 192-193)

27. What modus promotes the goals of a decentralized & bioregionally based food system?
    a) alternative livestock
    b) agroforestry
    c) organic farming
    d) biotechnology
    e) game farming and ranching
(p. 195)

28. Biotechnology's "*Bt*" found in cereals, pancake mixes, corn chips, & soy formulas is:
    a) a collection of *bug thorax* that are dehydrated during the packaging process
    b) a genetic mutation in *brown thyme* resulting from pesticide use
    c) a gene extracted from a soil bacterium (*Bascillus thuringiensis*)
    d) a *bagged termite* that has tunneled its way into the product
    e) of no risk to living organisms what so ever
(p. 197)

29. The Prairie Care Project helps conserve and restore wetland habitat for:
    a) a decrease in the number of farms and an increase in their size
    b) farmers as they try to pay off equipment at reduced profitability
    c) greater efficiency in the production of food crops and livestock
    d) waterfowl and other species
    e) CORE
(p. 200)

CHAPTER 7—FRESH WATER
(pp. 206-249)

STUDENT LEARNING OBJECTIVES

After studying this chapter you should be able to:
- understand the nature and distribution of Canada's freshwater resources
- identify a range of human uses of freshwater resources
- describe the impacts of human activities on fresh water and freshwater environments
- appreciate the complexity and interrelatedness of freshwater environment issues
- outline Canadian and international responses to freshwater issues
- discuss challenges to a sustainable future for freshwater resources in Canada
- define and discuss all of the KEY TERMS listed below

KEY TERMS (defined at the page number shown and in the glossary)

NOTES

Introduction
"Water is a precious and finite natural resource, yet water is often wasted and degraded."

Water Supply and Distribution

Earth's freshwater resources *(see Table 7-1)*

Source	% total resource	% fresh water resource
Ocean		
Ice caps/glaciers		
Land		
Atmosphere		
Human-made stores		

Canada's freshwater resources

Water Uses and Pressures on Water Quantity

Water uses

Term	Explanation and examples
Instream uses	
Withdrawal uses (see Figure 7-2, Figure 7-3, and Table 7-3)	
❑ *Intake*	
❑ *Discharge*	
❑ *Consumption*	
❑ *Recirculation*	
❑ *Gross water use*	
Point sources	
Nonpoint sources	

Balancing between withdrawal and instream uses

Pressures in Water Quality
Management of water quality (four aspects)

Biological Oxygen Demand (BOD)

The Importance of Water
Water as a common link

Water as a source of conflict: Most recently "water terror" has been an effective weapon in civil wars around the world.

Water as a hazard

Human Activities & Impacts on Freshwater Environments

Domestic and urban uses and impacts

Safe drinking water & sanitation facilities	*Drinking water guidelines & Regulations*

Drinking water treatment facilities

Typical municipal water treatment system (see Figure 7-4)

Grading Canada's drinking water treatment systems (*Table 7-4*)

Wastewater treatment facilities (Figure 7-5)

Primary treatment	Secondary treatment	Tertiary treatment

*Typical sewage treatment plant and grading wastewater treatment
(see Figure 7-5, Table 7-5)*

Demand for water (see Figure 7-6)

Pollution terminology

Persistence	Biocaccumulation	Biomagnification

Critical Pollutants in the Great Lakes Basin Ecosystem (Table 7-6)

Pollutant	Use	Method of entry

Industrial uses and impacts

Groundwater contamination

Impacts on Beluga Whales

Acidic deposition

Synergistic effects

Hydroelectric generation and impacts

Selected ecological and social impacts of dams and diversions (Table 7-7)

Agent	Process/Comment	Impacts

Hydroelectric Dams in Northern Quebec (Enviro-Focus 7)

Recreational uses and impacts

Responses to Environmental Impacts and Change

International initiatives

Agenda 21	
The Ramsar Convention	

Agreements between Canada and the United States

The Great Lakes Water Quality Agreements	
Remedial Action Plans	

Canadian Law, Policy, and Practice
Water legislation and policy responses

Inquiry on Federal Water Policy

Shared jurisdiction

Financial constraints

Research and application

Ecological Monitoring and Assessment Network	
Great Lakes Cleanup Fund and Great Lakes 2000 "Enviropark"	
Northern River Basins Study	

Canadian partnerships and local action
Flood damage reduction program (see also Figures 7-7 and 7-8)

Watershed planning

Fraser River Action

North American Waterfowl Management Plan

Making a difference locally (see also Box 7-5)

Future challenges

TRACKING YOUR PROGRESS
Chapter 7—Fresh Water

Multiple Choice

1. Though a reasonable quality of life is sustained on 80 litres per day, Canadians use
_____ litres per day.
   a) 50
   b) 95
   c) 175
   d) 340
   e) 745
(p. 207)

2. What is the amount of "fresh water" available on land from the Earth's total water supply?
   a) 0.61%
   b) 1%
   c) 3.76%
   d) 86%
   e) 97.8%
(p. 208)

3. Groundwater is defined as:
   a) aquifers
   b) all subsurface water
   c) submerging streams
   d) wells
   e) quarries
(p. 209)

4. Water supply per capita (m3/year) for Canada is _____ and for India is _____ .
   a) 120, 11.5
   b) 45.2, 31.19
   c) 3.19, 1.9
   d) 19.7, 2.23
   e) 120, 1.9
(p. 210)

5. The Canadian region with the least amount of average annual precipitation is:
    a) Atlantic
    b) Pacific
    c) Gulf of Mexico
    d) Hudson Bay
    e) Atlantic
(Pp. 210-211)

6. Approximately _____ per cent of Canada's water drains SOUTH:
    a) 90
    b) 70
    c) 40
    d) 20
    e) 12
(p. 210)

7. What major flood was attributed to intense rainfall?
    a) Saguenay River-Lac Saint Jean, 1996
    b) Oldman and South Saskatchewan Rivers, 1959
    c) Lake Baikal and adjacent lakes in the former USSR, 2000
    d) Lake Victoria and adjacent lakes in Africa, 1809
    e) Rice Lake and adjacent watersheds, 1987
(p. 210)

8. Which of the following is NOT a point source pathway?
    a) pipe
    b) ditch
    c) channel
    d) agricultural runoff
    e) conduit
(p. 212)

9. What system constitutes the largest water withdrawal use in Canada (1972-1991)?
    a) mining
    b) thermal power
    c) agriculture
    d) manufacturing
    e) municipal
(p. 212)

10. Which province had the largest per capita urban water use in 1996?
     a) Ontario
     b) Newfoundland
     c) Prince Edward Island
     d) British Columbia
     e) Quebec
(p. 213)

11. What was the revenue generated by Canadian sport fishers in 1996?
     a) $5.9 billion
     b) $14 million
     c) $1.2 billion
     d) $2.4 billion
     e) $96 million
(p. 214)

12. Agricultural uses raise many questions but NOT the following:
     a) thermal pollution
     b) BOD
     c) soil salinization
     d) erosion problems
     e) none of the above
(p. 215)

13. Commercial navigation requires high water levels which may cause_____.
     a) bank erosion
     b) disruption of bottom sediment
     c) beach ecosystems to change
     d) degradation of water quality when dredged to maintain depth
     e) all of the above
(p. 215)

14. Following the accidental introduction of zebra mussels, this did NOT occur:
     a) water became less turbid (clearer) as a result of their feeding
     b) phytoplankton food source decreased
     c) the concentration of particulates were filtered out
     d) industrial intake pipes unclogged due to their voracious feeding habits
     e) machinery jammed
(Pp. 215-216)

15. Ransom (1995) reported that to discharge pollutants into water is akin to allowing:
    a) evil spirits to enter the body
    b) the disconnection of the mental from the physical
    c) a physical essence to drive evil spirits into action
    d) substances to disconnect the beauty of the natural world
    e) drugs to be injected into [your] mother's blood
(p. 217)

16. The 1997 Red River flood inundated some Manitoba communities despite _____.
    a) damming
    b) evacuation
    c) pumping
    d) sandbagging
    e) raised bungalow complexes
(Pp. 218-219)

17. The May 2000 Walkerton Ontario water treatment tragedy was caused by:
    a) *Echericoda boli* 0157:H7 contamination
    b) *Esthericdo cobi* 175:7H contamination
    c) *Escherichia coli* 0157:H7 contamination
    d) *Elephantitis HIV* 0175:H77 contamination
    e) *Erichdicado candie* 09453:MZ contamination
(p. 220)

18. Where in Canada can you find the best drinking water?
    a) Ontario
    b) Alberta
    c) Quebec
    d) (a) and (c) only
    e) (b) and (c) only
(p. 222)

19. Typically, secondary treatment of municipal sewage involves:
    a) a chemical process to remove nutrients (phosphates, nitrates) and metals
    b) degradation of dissolved organics by microorganisms
    c) mechanical removal of solids, sediment, and some organic matter
    d) removal salts, acids and organochlorine compounds
    e) all of the above
(p. 222)

20. Vacuum filter or centrifuge systems remove water from processed sludge in the:
    a) dewatering process
    b) primary settling
    c) aeration tanks
    d) primary digest
    e) concentrated sludge
(Pp. 222-223)

21. The worst wastewater treatment systems in Canada as presented in Draper are found in:
    a) Vancouver, Brandon, Ottawa, and Fredericton
    b) Victoria, Dawson Saint John, and Halifax
    c) St. John's, Montreal, and Charlottetown
    d) Dawson City, St. John's, Montreal, and Victoria
    e) Calgary, Toronto, Yellowknife, and Edmonton
(Pp. 222-224)

22. In a Canadian home, most fresh water use is attributed to:
    a) cleaning
    b) laundry
    c) baths and showers
    d) toilets
    e) drinking and kitchen use
(Pp. 225-226)

23. DDT and PCBs accumulating and concentrating in the fatty tissue of predators is called:
    a) persisting
    b) bioaccumulating
    c) biomagnifying
    d) registering
    e) regenerating
(Pp. 227-228)

24. DNAPLs are NOT:
    a) nonaqueous phase liquids
    b) cleaning solvents
    c) wood rotting solvents
    d) used in asphalt operations
    e) used in automobile production
(p. 229)

25. Montreal researchers discovered that 14 of 73 Beluga cadavers suffered from:
    a) bone disease
    b) hypothermia (due to lack of body fat)
    c) mercury poisoning
    d) exposure to PCBs
    e) cancer
(p. 231)

26. Canada's and the United States' response to human impacts on water was the:
    a) Air Quality Agreement
    b) Nonaqueous Substance Act
    c) Clean Water Act
    d) Experimental Lakes Area
    e) Schindler and Bayley references
(p. 232)

27. Impounding water in a reservoir does NOT:
    a) displace people
    b) disrupt wildlife
    c) influence local climate
    d) increase dissolve oxygen levels
    e) cause small earthquakes
(Pp. 232-233)

28. An example of international initiative(s) to protect the quality freshwater includes:
    (a) Agenda 12
    (b) Ramsar Convention
    (c) World Water Ways
    (d) (a) and (b) only
    (e) all of the above
(Pp. 236-237)

29. What Environment Canada (1994) initiative was designed to improve communication and co-operation among scientists conducting ecological monitoring activities?
    a) RAP
    b) CCWI
    c) NWRI
    d) PCB
    e) EMAN
(Pp. 239-240)

CHAPTER 8—OCEANS AND FISHERIES
(pp. 250-294)

STUDENT LEARNING OBJECTIVES

After studying this chapter you should be able to:
- identify a range of human activities occurring in the marine environment
- discuss the effects of human activities on the marine environment
- describe the complexity and interrelatedness of marine environment issues
- specify Canadian and international responses to oceans and fisheries issues
- discuss challenges to a sustainable future for oceans and fisheries resources
- define and discuss all of the KEY TERMS listed below

KEY TERMS (defined at the page number shown and in the glossary)

NOTES

Introduction

Exclusive Economic Zone (EEZ) (see also Figure 8-1a)

Canada's Arctic, Pacific, and Atlantic Ocean environments
Major Characteristics

Arctic Ocean Environment	Pacific Ocean Environment	Atlantic Ocean Environment
Major Characteristics	*Major Characteristics*	*Major Characteristics*
• ocean area	• ocean area	• ocean area
• polynyas		
• sustainable yield		
• bioaccumulates	*Threats to ecosystem integrity*	*Threats to ecosystem integrity*
	• global change	• lack of knowledge
• biomagnification		
Threats to ecosystem integrity		
• hydroelectric projects		• anthropogenic impacts & marine pollution
• LRT of pollutants	• marine pollution	
• climate change		• commercial fishing
• overhunting		
• nonrenewable resource extraction		• sea-level rise

Summary of concerns facing Canada's ocean regions

World's most-fished species (Figure 8-3a)

National percentage of total world fish catch, 1989 (Figure 8-3b)

World total fish production in marine waters, 1950-1993(Figure 8-3c)

Human Activities and Impacts on Marine Environments

Fisheries

Human Activities and Impacts on Marine Environments (continued...)
The 1995 Canada-Spain Turbot Dispute

50 million meals dumped at sea (Enviro-Focus 8)

The Northern Cod Moratorium (see also Table 8-1)

Pacific Herring and Salmon Stocks

Pacific Herring

Coastal Salmon

Pollution

Industrial and Chemical Effluents (see also Table 8-2)

Municipal sewage

Marine shipping, ocean dumping, and plastics
Marine shipping

Aquaculture

Ocean dumping

Plastics

Coastal development
Urban runoff

Physical alterations

Offshore hydrocarbon development

Atmospheric change

Responses to Environmental Impacts and Change

International initiatives

United Nations Convention and the Law of the Sea (see also Figure 8-12)

Agenda 21

Canadian law, policy, and practice
Revised oceans and fisheries legislation (see also Table 8-4)

Coastal Zone Management (EEZ) efforts

Canadian partnerships and local actions

Future challenges

TRACKING YOUR PROGRESS
Chapter 8—Oceans and Fisheries

Multiple Choice:

1. What are the local Arctic currents that help maintain open water areas?
    a) phytoplankton
    b) polynyas
    c) circular fronts
    d) polar vortex
    e) walrus rings
(p. 253)

2. The long and complex food web that supports top predators in the Arctic has _____ tropic levels.
    a) 2
    b) 3
    c) 4
    d) 5
    e) 6
(p. 253)

3. The most import groups of Long-Range Transport Pollutants in the Canadian Arctic are:
    a) clams and oysters
    b) PBCs and DTD
    c) organic pollutants and heavy metals
    d) snow and rain
    e) whales and seals
(p. 255)

4. Due primarily to sewage contamination, this was closed in Boundary Bay:
    a) salmon fisheries
    b) cod fisheries
    c) squid harvesting
    d) shellfish harvesting
    e) there are NO closures in effect in BC as it has the best water treatments plants on Earth
(p. 256)

5. Which of the following does NOT influence marine life diversity (Atlantic)?
    a) melting ice from the high Arctic
    b) local gyres
    c) eddies
    d) tides
    e) spring discharge from northern rivers
(p. 257)

6. Which of the following is NOT one of the most-fished species in the world?
    a) Peruvian Anchoveta
    b) Japanese Pichard
    c) Alaskan Pollock
    d) Atlantic Salmon
    e) Chilean Jack Mackeral
(Pp. 259-260)

7. The FAO has estimated that _____ per cent of marine species have reached or
    exceeded sustainable levels, and adopted the formal, global code of conduct in 1995.
    a) 90
    b) 85
    c) 70
    d) 40
    e) 20
(p. 259)

8. The EEZ, an exclusive jurisdiction over a shore zone, extends out _____.
    a) 20 nautical miles
    b) 65 nautical miles
    c) 120 nautical miles
    d) 150 nautical miles
    e) 200 nautical miles
(p. 262)

9. The Canadian-Spain conflict revolved around what species?
    a) *Bufo americanus*
    b) *Reinhardtius hippoglossoides*
    c) *Rana sylvatica*
    d) *Pincea glaucus*
    e) *Homo sapiens*
(p. 262)

10. The turbot hostilities with Spain (1995) cost Canadian taxpayers over _____ .
    a) $75 000
    b) $3 million
    c) $200 million
    d) nothing, the Canadian government paid all the costs
    e) nothing, the Spanish government paid all the costs
(p. 264)

11. How much fish was wasted fish off the Alaskan coasts from 1992 to1994?
    a) 2000 kg
    b) 340 million kg
    c) 903 million kg
    d) 1.7 billion kg
    e) 30 tonnes
(Pp. 265-266)

12. What were the impacts of the 1968 northern cod catch, labelled "the killer spike"?
    a) the destruction of new technology and huge factory ships
    b) the inappropriate politically-motivated decisions to increase fishing in 1969
    c) an increase in use of trawling nets and spear fishing
    d) a removal of Newfoundland's inshore cod population and stock resiliency
    e) an abundant east coast fishery and prosperous Newfoundlanders
(p. 265)

13. Which of the following was NOT a reason for the fall of the groundfish fishery in Atlantic Canada?
    a) corporate interests
    b) driftnetting
    c) ghostfiling
    d) predation by seals
    e) government policies
(p. 268)

14. Which of the following is a threat to the survival of Pacific herring?
    a) coastal development
    b) Pacific hake
    c) abundant kelp beds
    d) uncontaminated waters
    e) both (a) and (b) are correct
(p. 269)

15. "The force of economics overpowers conscience" (Parfit, 1995) refers which of the following fishery?
    a) Pacific herring
    b) Chinook salmon
    c) Coho salmon
    d) Pacific salmon
    e) Atlantic herring

(Pp. 272-273)

16. Contamination of Canada's oceans and coastlines is principally the result of:
    a) long-range transport of pollutants from overseas
    b) tides
    c) oceanic drifts from the USA
    d) acid rain
    e) human activities

(p. 274)

17. Which of the following is NOT a nonpoint source of pollution?
    a) municipal sewage
    b) urban runoff
    c) agricultural runoff
    d) acid rain
    e) car exhaust

(p. 274)

18. The largest fine paid to date for pollution violation under the federal Fisheries Act is:
    a) $150
    b) $23 000
    c) $250 000
    d) $500 000
    e) $750 000

(p. 275)

19. Who said that heavy metals in stream sediments had increased in ten years? (Yukon)
    a) Environment Canada, from 1985 to 1995
    b) Natural Resource, from 1987 to 1997
    c) Keno Hill Mine Annual Self-Reporting, from 1989 to 1999
    d) Little River Inc., from 1990 to 2000
    e) Domtar Specialty, from 1991 to 2001

(p. 275)

20. An Inuk mother attempted to protect her baby from PCB contamination by feeding her:
    a) breast milk
    b) country food
    c) Carnation powdered milk
    d) Coffee Mate mixed with water
    e) Cream of Wheat
(p. 276)

21. These municipalities have been dumping raw sewage into the Atlantic for 250 years:
    a) Saint John and Quebec City
    b) Vancouver and Victoria
    c) Halifax and Dartmouth
    d) St. John's and Montreal
    e) Inuvik and Fairbanks
(p. 278)

22. Marine birds and other creatures are hurt or killed when they mistakenly eat:
    a) fish
    b) plastics
    c) shells
    d) herring waste
    e) sea weed
(p. 280)

23. The Federal Court of Canada concluded that the Confederation bridge:
    a) posed no harm to the environment
    b) will cause irreparable harm to coastal ecosystems
    c) was ill-planned and showed structural deficiencies
    d) would trap whales causing them to be beached and die
    e) had not been given a fair chance before it was dismantled and sold in parts
(Pp. 281-282)

24. What event began in 1973 off the Arctic coast?
    a) search for thermal vents
    b) search for fresh water stores under the ocean floor
    c) DDT and other polychlorinated biphenyls monitoring
    d) monitoring for locked heavy metal burdens
    e) exploration for petroleum hydrocarbons
(p. 283)

25. The international basis on which to pursue protection and sustainable development of marine and coastal environments and their resources is _____ .
     a) UNCLOS
     b) UNEP
     c) *Agenda 21*
     d) EEZs
     e) FAO
(p. 286)

26. Which of following is NOT a Federal Department with significant marine responsibilities?
     a) Environment
     b) Consumer Affairs and Business Development
     c) Indian Affairs and Northern Development
     d) Foreign Affairs and International Trade
     e) Emergency Preparedness Canada
(p. 286)

27. What was the stimulus for the creation of the Atlantic Coastal Action Program?
     a) recognition of larger government's role
     b) growing demand for public involvement in decision making
     c) coordinate policies and programs
     d) all of the above
     e) none of the above
(p 288)

28. The enthusiasm of local communities toward the ACAP process was witnessed by:
     a) rapid start-up of initiatives
     b) little pollution of the impounded water
     c) little impact on the flooded land
     d) reduced levels of BOD and TSS in the impounded water
     e) higher species diversity in the sixth year
(p. 288)

29. The management of oceans and fisheries resources for future generations requires:
     a) particular fish stocks
     b) high species diversity and cold temperatures
     c) stewardship and a shared awareness of both ethical and ecological principles
     d) animal populations that can not be easily be eliminated
     e) knowledge-building activities
(Pp. 289-290)

CHAPTER 9—FORESTS
(pp. 296-347)

STUDENT LEARNING OBJECTIVES

After studying this chapter you should be able to:
- understand the nature and distribution of Canada's forest resources.
- identify a range of human uses of forest resources.
- describe the impacts of human activities on forests and forest environments.
- appreciate the complexity and interrelatedness of forest environment issues.
- outline Canadian and international responses to forest issues.
- discuss challenges to a sustainable future for forest resources in Canada.
- define and discuss all of the KEY TERMS listed below.

KEY TERMS (defined at the page number shown and in the glossary)

clear-cutting	...307	highgrading	...307
continuous clear-cutting	...308	silviculture	...309
deforestation	...310	sustained yield	...310

NOTES

Introduction
The boreal forest that drapes "like a great green scarf across the shoulders of North America", continues to enrich the lives of all Canadians.

Focal Point: Clayoquot Sound

Selected statistics (see Figure 9-1)	
First Nation people	
Land area	
Commercially productive forest land	
Area logged by 1995	
Unlogged merchantable forest	
Protected areas	
Primary or old-growth forest	

Focal Point: Clayoquot Sound (continued...)

The Earth's Forests
Global distribution, products, and demand (see also Figure 9-2)

Forests in Canada *(see also Figure 9-3)*

Major Forest Region	Comments
Boreal *1. Predominantly forest* *2. Forest and barren* *3. Forest and grass*	
Subalpine	
Montane	
Coast	
Columbian	
Deciduous	
Great-Lakes *St. Lawrence*	

Harvesting systems (*see also Figure 9-5 and Table 9-3*)

Harvesting systems	Sketch	Comments
Selective (or selection) cutting		
Shelterwood cutting		
Seed tree cutting		
Clear-cutting		
Patch (clear) cutting		
Strip cutting		
Whole tree harvesting		

Tree plantation - silviculture

Deforestation

The timber bias (see also Figure 9-6)

The falldown effect (see also Figure 9-7)

The Ecological Importance of Old-Growth Forests

The Life Cycle in the Old-Growth Forest
Standing live trees, snags, and fallen trees

Carbon storage

Keys to diversity

Biological diversity

The need for protection

Human activities and impacts on forest environments

A brief historical overview of the forest industry
First Nations and European Settlers

Timber exports and government ownership

A brief historical overview of the forest industry (continued...)
Changing market demands, changing industry

The shifting frontier and conservation concerns

Licensing of Forest Holdings

Growth in pulp and paper and increased concentration

Sustained-yield focus

Timber production activities and their effects

Habitat, wildlife, and life-support effects (Table 9-4)

Examples of forest-dwelling species at risk in Canada (1995)			
	Mammals	Birds	Plants	Reptile
Endangered				
Threatened				
Special concern				

Degradation and deforestation of tropical forests

Pollution

Sociocultural dimensions

Tourism and recreation

Human activities and Ontario's Carolinian forest

Responses to Environmental Impacts and Change

International initiatives
UNCED Forest Principles

Agenda 21

UNCED Conventions and other responses

Canadian policy, practice, and partnerships *(see also Table 9-6)*

Canada's National Forest Strategy

Canada's Model Forest Program

Criteria and indicators of sustainable forest management (see Table 9-7)

Local partnership and responses (see also Box 9-6)

Wildlife and Forestry Activity, Eco-adventures, and BC's Forest Practices code

Future challenges

TRACKING YOUR PROGRESS
Chapter 9—Forests

Multiple Choice Questions

1. What has presented a big forestry-type challenge to countries in the last 20 to 30 years?
   a) old growth and cutting access
   b) technology and cutting methodologies
   c) environmental, economic, social and political dimensions
   d) deep ecologists living high in the trees in protest of cuts
   e) wet climate not allowing cutters to do their job

(p. 297)

2. Where would one find some long-lived forest trees, with ages greater than 1000 years?
   a) Pacific Northwest
   b) Arctic South
   c) Atlantic Southeast
   d) Pacific Southeast
   e) Atlantic Northeast

(p. 298)

3. The citizens of Clayoquot Sound protested the decision to allow major clear-cuts (+2/3) in _____.
   a) boreal forests (1999)
   b) rain forests (1993)
   c) deciduous forests (2000)
   d) alpine forests (1898)
   e) transitional forests (1927)

(p. 299)

4. The compromise which resulted from the conflict in Question 3, was the formation of a:
   a) Forestry Panel
   b) Fisheries Panel
   c) Wildlife Panel
   d) Rain Forest Panel
   e) Scientific Panel

(p. 299)

5. The coastal temperate rain forests found in the British Columbia and Alaska are:
    a) globally important
    b) commonly found world wide and thriving
    c) include Sitka spruce, Douglas fir, Western hemlock and Yellow cedars
    d) (a) and (b) only
    e) (a) and (c) only
(p. 300)

6. Threats to forests in developed and developing countries do NOT include:
    a) fire
    b) pollution
    c) clearing for agriculture
    d) fuelwood
    e) none of the above
(p. 301)

7. In Canada there are _____ type(s) of Boreal forests.
    a) 1
    b) 2
    c) 3
    d) 4
    e) 5
(Pp. 302-303)

8. In Canada, forest management is a matter of _____ jurisdiction.
    a) federal
    b) provincial
    c) regional
    d) municipal
    e) private
(p. 303)

9. How many forest hectares were cleared for settlement in North America between 1860 and 1978?
    a) 12 thousand
    b) 89 thousand
    c) 3 million
    d) 64 million
    e) 19 billion
(p. 305)

10. Approximately how many tree species are found in Canada?
    a) 47
    b) 79
    c) 180
    d) 1 278
    e) 15 900
(p. 304)

11. Which pulp & paper mill turns about 30 football fields of forest to pulp every day?
    a) Idaho's Boise Cascade
    b) Quebec's Noranda mines
    c) Ontario's Quaker Oaks
    d) Saskatchewan's Prince Albert pulp mill
    e) Alberta's Pacific pulp mill
(p. 306)

12. Which of the following forest harvesting method leaves wooded corridors (may serve as seed source)?
    a) selective cutting
    b) shelterwood cutting
    c) strip cutting
    d) patch cutting
    e) clear-cutting
(Pp. 307-308)

13. Which country is referred to as the "Brazil of the North"?
    a) Canada
    b) USA
    c) Mexico
    d) Greenland
    e) Russia
(p. 310)

14. What is used to describe the potential of ecological maturity versus economic maturity?
    a) black boards
    b) loud speakers
    c) forests and cities
    d) fish and rivers
    e) green and ripe tomatoes
(Pp. 311-312)

15. What will occur if we shift from exploiting old-growth forests to exploiting managed forests?
    a) without a doubt, the smashed green tomato effect
    b) the falldown effect
    c) the alchemy of fibre
    d) a cornerstone production
    e) an undesirable monoculture
(Pp. 312-313)

16. Old growth wood is NOT _____.
    a) clear
    b) soft
    c) knotty
    d) fine-grained
    e) easy to work with
(p. 313)

17. Old-growth forests can feed a diverse sustainable industry including:
    a) large and small log sawmills
    b) pulp mills, beam and truss lamination plants
    c) cabinet shops and furniture factories
    d) millwork plants
    e) all of the above
(p. 314)

18. The costs of "forestopia" will have to be paid by _____ .
    a) its falldown
    b) monoculture practices
    c) a fibre economy
    d) revenues generated from hope-for value-added activity
    e) the new woodlots created
(p. 315)

19. How many mycorrhizal fungi species can grow on roots of healthy old-growth trees?
    a) 30 to 40
    b) 100 to 250
    c) 400 to 750
    d) 3 to 6
    e) unknown
(Pp. 315-316)

20. The quote, "Nature creates forests – we can only watch" is from:
    a) Marchak, 1995
    b) Schoonmaker, 1997
    c) von Hagen, 1997
    d) Parfitt, 1994
    e) Hammond, 1991
(Pp. 315-316)

21. "A wide range of services, products, and values" is:
    a) Boise Cascade Pulp and Paper's motto for happy customers
    b) what Canadians depend on
    c) what indicates "best next tree-crop rotation procedures"
    d) what nature depends on
    e) the motto for highgrading forest products
(p. 319)

22. Indigenous people probably used clean mosses from the forest floor to make _____.
    a) medicine
    b) tobacco
    c) disposable diapers (today derived from wood pulp; this we often forget)
    d) a type of investment for trade
    e) a soup supplement
(p. 319)

23. From the 17th to the 19th centuries, the vast majority of European immigrants wanted to:
    a) have vacations in Canada
    b) establish farms in Canada
    c) preserve the wildlife found in Canada
    d) conserve the forests of Canada
    e) return to Europe from Canada
(p. 319)

24. John Langton warned people of the late 19th century that Canadian forests were:
    a) about to be exploited by Americans
    b) the property of the European state
    c) a source of current revenue not a right of the people
    d) finite
    e) infinite
(p. 320)

25. BC's "Tree Farm Licenses" (1945) were established in law and initially granted for:
    a) 25 years
    b) 45 years
    c) 60 years
    d) 75 years
    e) perpetual terms
(p. 321)

26. What is the period of time that clearcuts degrade or destroy most non-timber uses and values?
    a) one or more human generations
    b) 40 years
    c) no time as this does not happen
    d) 25 years
    e) clear-cuts are great such that non-timber uses and values triple in 5 years
(p. 320)

27. Pallid bat, blue ash, Yellow-breasted chat, and the Black rat snake are:
    a) endangered species
    b) threatened species
    c) vulnerable species
    d) rare species
    e) extinct species
(Pp. 322-324)

28. As trees disappear through logging so may _____ .
    a) woodland caribou
    b) fish
    c) coral reefs
    d) plants
    e) all of the above
(p. 322)

29. Tropical forests do NOT produce _____ .
    a) nuts
    b) fruits
    c) chocolate
    d) vinyl
    e) rubber
(p. 325)

30. This BC mushroom is one of 30 species picked commercially on a large scale:
    a) pine mushroom, a green jewel
    b) pine mushroom, a rose nut
    c) pine mushroom, a base ball
    d) pine mushroom, a fussy fungi
    e) pine mushroom, a poof-poof
(p. 327)

31. Caribou live in the Canadian _____.
    a) temperate rain forests
    b) boreal forests
    c) tundra
    d) alpine forests
    e) prairies
(p. 328)

32. Minimata disease comes from _____.
    a) arsenic in caribou meat contaminated by gold mine effluents
    b) lead in *mycorrhizal fungi* contaminated by forestry operators
    c) mercury contaminated fish from the Ontario Wabigoon River system
    d) the Far East transported to Canada by birds
    e) Martian meteors mixing in agricultural soil, then taken up by crops
(p. 329)

33. What are the narrow bands of trees left to ring major lakes and streams used by tourists?
    a) strip cuts
    b) patch cuts
    c) view-scape cuts
    d) riparian cuts
    e) buffer cuts
(p. 331)

34. This 265 hectare forest is one of the best examples of a mature Carolinian forest in Canada.
    a) Rouge River Valley
    b) Backus Woods
    c) Jack Miner Bird Sanctuary
    d) Canada's Wonderland Park
    e) African Safari
(Pp. 332-333)

35. Which of the following is one of eight, non-legally binding UNCED forest principles?
    a) establish Ontario as the top world exporter of paper fibres
    b) include New Brunswick in the free-trade act
    c) encourage fair international trade in forest products
    d) perceive trees as giant vegetables potentially ready to harvest
    e) promote public harvesting of old, decaying, standing wood
(p. 334)

36. Why was 100 000 to 250 000 hectares of Canadian forests set aside in 1991?
    a) for clearcutting in the next two years
    b) for the model forest program
    c) for old-growth preservation
    d) for extremely rare critters
    e) for Canadians to do what they want and when they want
(p. 336)

37. What does the term "utilization and rejuvenation are balanced and sustained" stand for?
    a) conservation of biological diversity
    b) maintenance and enhancement of forest ecosystem condition and productivity
    c) conservation of soil and water resources
    d) forest ecosystem contributions to global ecological cycles
    e) socioeconomics
(p. 338)

38. This involves various kinds of formal and informal partnerships with forest firms, local community groups, and environmental non-government organizations (ENGOs)
    a) NGOs
    b) Agenda 21
    c) CCFM
    d) ISO 14 000
    e) Canadian Council of Forest Ministers
(Pp. 337-341)

39. One way to protect forests and forest ecosystems is to identify _____.
    a) fast growing trees
    b) vacant lands for tree planting
    c) alternative source of fibre, such as hemp
    d) who is destroying the forest
    e) how forest ecosystems shrink
(Pp. 341-342)

CHAPTER 10—MINING
(pp. 348-378)

STUDENT LEARNING OBJECTIVES

After studying this chapter you should be able to:
- understand the nature and distribution of Canada's mineral resources
- describe the impacts of mining on natural environments
- outline Canadian and international responses to mining issues
- discuss challenges to sustainable mining in Canada
- define and discuss all of the KEY TERMS listed below

KEY TERMS (defined at the page number shown and in the glossary)

NOTES

Introduction
Is mining something we could readily choose not to do?

Human Activities and Impacts on Natural Environments

Historical overview

One-industry towns

Long-distance commuting

Production, value, and distribution of mineral resources in Canada
(see also Table 10-1)

Mineral fuels	Nonfuel minerals	Precious metals	Structural Materials

Canada's first diamond mine: As delightful as beautiful diamonds may be to look at, they are a luxury; they are not necessary to live a good and happy life.

Environmental impacts of mining

Acid mine drainage

Mineral exploration

Potential environmental impacts of mining (see also Table 10-2 and Figure 10-4)

Environmental impact assessment (EIA) process (see also Table 10-3)

Environmental impact statement (EIS) (see also Table 10-3)

Mine development and mineral extraction

Surface mining techniques

Open pit	Strip mining	Overburden

Processing of minerals

Milling	Tailings

Mine closure and reclamation: Abandoned mine and orphaned mines (see Table 10-4)

Canada's MEND program (see Box 10-2)

Responses to environmental impacts and change
Market forces

Partnerships for environmental sustainability

Future Challenges
Stewardship

Protection and monitoring

Knowledge building

TRACKING YOUR PROGRESS
Chapter 10—Mining

Multiple Choice Questions

1. How many Canadians were employed in the mining and mineral processing industries in 1999?
   a) 3 600
   b) 78 000
   c) 83 000
   d) 123 000
   e) 386 000
(p. 349)

2. What is the imbalance between the distribution of people and natural resources in Canada?
   a) gold rush
   b) tip-topping
   c) single resource towns
   d) one-industry towns
   e) both (c) and (d) are correct
(p. 350)

3. This is the term used when miners fly into a mine to work for a designated period and are flown back home.
   a) long-distance commuting
   b) piggy-backing
   c) kiss and fly
   d) homing pigeons
   e) there's no place like home
(p. 351)

4. Non-fuel minerals do NOT include which of the following?
   a) iron ore
   b) coal
   c) gold
   d) silver
   e) copper
(p. 351)

5. Which is the largest mining project of the NWT & the first diamond mine in North America?
   a) Slave Geological Province (SGP)
   b) Potash Proprietary (PP)
   c) Broken Hill Proprietary (BHP)
   d) Charles Fipke and Stewart Blusson (CFSB)
   e) Dia Met Minerals (DMM)
(p. 354)

6. Treaty 11 Dogrib, requires BHP to develop monitoring and management plans for:
   a) bison and penguins
   b) birds and bears
   c) amphibians and reptiles
   d) caribou and birds
   e) all of the above
(Pp. 354-355)

7. What was created to protect the tundra hills and calving grounds of Bluenose caribou?
   a) Tuktut Nogait National Park in 1996
   b) World Wildlife Fund in 1996
   c) EARP in 1996
   d) Panda National Park in 1996
   e) Norman Wells Pipeline in 1996
(p. 355)

8. Yellowknife's Dene are concerned about the fuel oil, arsenic, and cyanide that _____.
   a) are smelted and refined
   b) will be mined and milled
   c) BHP will extract and process
   d) will be hauled on winter roads across their hunting grounds
   e) leaked into the permafrost
(p. 356)

9. What is the most serious environmental problem facing the mining industry?
   a) basic rock erosion
   b) acid mine drainage
   c) pyrites steeling the diamonds
   d) rusting oxides
   e) oxidizing diamonds
(Pp. 357-358)

10. The following are hazards linked with the mining process & the environment:
    a) land degradation
    b) ecosystem disruption
    c) metal particles
    d) chemical leakage
    e) all of the above
(p. 359)

11. Which of the following identifies consequences of undertaking new developments and changing natural systems?
    a) IES
    b) EIA
    c) AIE
    d) IAE
    e) ESI
(Pp. 360-361)

12. What sort of mining is used when material lies nearly flat and is NOT buried too deeply?
    a) open pit mining
    b) scoop-a-pit
    c) strip mining
    d) crop mining
    e) overburden dig
(p. 362)

13. Extensive surface excavations of overburden materials may _____ .
    a) clog streams
    b) create excessive dust
    c) disturb habitat
    d) (a), (b) and (c) are correct
    e) none of the above
(p. 362)

14. These are used to remove explosive and radioactive gases, and to provide fresh air.
    a) pump sumps
    b) usually a second shaft
    c) sophisticated ventilation systems
    d) both (b) and (c) are correct
    e) all of the above are correct
(p. 362)

15. What chemical is NOT used in the flotation process of base milling?
    a) formaldehyde
    b) kerosene
    c) organic agents
    d) sulphuric acid
    e) none of the above are used
(p. 363)

16. Which of the following are all base metals?
    a) copper, iron, and zinc
    b) steel, aluminium, and nitrogen
    c) copper, lead, zinc, and nickel
    d) nickel, zinc, and steel
    e) aluminium, nitrogen, and copper
(p. 364)

17. The smelter producing the most sulphur dioxide emissions in 1994 was _____ .
    a) Inco Ltd., Sudbury, ON
    b) Inco Ltd., Thompson, MB
    c) Canadian Electrolytic Zinc, Valleyfield, QC
    d) Noranda Minerals Inc., Murdochville, QC
    e) Hudson Bay Mining and Smelting, Flin Flon, MB
(Pp. 364-365)

18. In the past, little if any concern was shown for _____ .
    a) land restoration
    b) mineral extraction
    c) environmental stability
    d) public safety
    e) all of the above, but NOT (b)
(p. 365)

19. Today mines have few downstream impacts because they are no longer _____ .
    a) next to rivers
    b) built over grades
    c) orphaned
    d) closed
    e) allowed
(p. 366)

CHAPTER 11—ENERGY
(pp. 380-411)

STUDENT LEARNING OBJECTIVES

After studying this chapter you should be able to:
- understand the supply and demand of energy resources in Canada
- identify a range of human uses of energy resources
- describe the impacts of human activities related to the production and use of Canada's energy resources
- appreciate the complexity and interrelatedness of energy issues
- outline Canadian and international responses to energy issues
- discuss challenges to sustainable energy production and use in Canada
- define and discuss all of the KEY WORDS listed below

KEY TERMS (defined at the page number shown and in the glossary)

NOTES

Introduction
Canadians and energy

Human Activities and impacts on Natural Environments
Energy supply and demand in Canada

How do we use energy?

Primary energy use	Secondary energy use	Petajoules

Net useful energy

Energy resources

Energy resources (continued...)

Fossil fuels

Type	Notes
Oil	
Heavy oil • *oil shale* • *oilsand* • *bitumen*	
Coal	
Natural gas	

The 1970s energy crises

Biomass

Hydroelectricity

Nuclear energy

Responses to Environmental Impacts and Change

Emerging energy resources
Alternative energy

Solar	Wind	Hydrogen

Barriers to adopting alternative technologies

Improving energy efficiency

Transportation efficiency	Industrial efficiency	Home efficiency

Organized initiatives

Energy futures

TRACKING YOUR PROGRESS
Chapter 11—Energy

Multiple Choice Questions

1. Which of the following is an example of "green" energy?
    a) natural gas
    b) coal
    c) wind
    d) petroleum
    e) oil
(p. 381)

2. What are WLED batteries charged by?
    a) pedal generators
    b) electricity
    c) nuclear energy
    d) wind
    e) sun
(p. 382)

3. Recently our energy use per dollar of GDP has been declining because:
    a) people would rather freeze than pay excessive heating bills
    b) of energy conservation practices
    c) Canadians can not afford the money required to heat at room temperature (18°C)
    d) more clothing is worn indoors today to help reduce the cost of heating
    e) Canadians are very friendly and rather huddle for warmth than pay for heat
(Pp. 382-383)

4. From 1990 to 1997, Canadian demand for primary energy use increased by 15.3 per cent, from 9500 to 10 955 _____.
    a) kilowatts
    b) calories
    c) volts
    d) petajoules
    e) watts
(p. 384)

5. When 35 units of coal energy produces 20 units of electricity, there is a net energy
_____ of _____ units over the lifetime of the system.
    a) gain; 15
    b) loss; 15
    c) gain; 55
    d) loss; 55
    e) gain; 105
(p. 385)

6. Unlike Hibernia, Terra Nova oil will be _____.
    a) produced from a concrete platform
    b) from limestone
    c) propane
    d) pumped from the seabed into a vessel capable of processing 14 000
        barrels/day
    e) generated from dead sea animals frozen in the high Arctic
(p. 385)

7. Hydrocarbons are NOT derived from the remains of prehistoric _____.
    a) animals
    b) forests
    c) fish
    d) corals
    e) meteors
(Pp. 385-388)

8. Where are the largest oil sand deposits in the world?
    a) the Sahara desert
    b) Egypt
    c) USA
    d) Canada
    e) Russia
(p. 388)

9. In order to reduce the risk of spills, tanker ships built after 1993 need to have _____.
    a) double hulls
    b) oil spill collection technology aboard the ship
    c) expensive transportation-oil escort ships (TOES)
    d) been made out of unsinkable materials
    e) an environmental spill technologists on board at all times
(p. 389)

10. What is the most abundant fossil fuel in the World?
     a) coal
     b) oil sand
     c) natural gas
     d) heavy oil
     e) crude oil
(p. 390)

11. Which fossil fuel is a gaseous mixture of methane, propane and butane?
     a) coal
     b) oil sand
     c) natural gas
     d) heavy oil
     e) crude oil
(p. 391)

12. Flaring produces 200+ chemical compounds, including 30+ varieties of
_____.
     a) intoxicating gases
     b) useable metals
     c) drinkable liquids
     d) poisonous fumes
     e) cancer-causing benzene
(p. 392)

13. When is biomass (wood for example) considered to be a renewable energy resource?
     a) only when prehistoric deposits are used
     b) when trees and plants are replaced at rates lower or equal to rates of harvest
     c) when only certain types of fast growing trees are used in the tropics
     d) only when used out of necessity
     e) when we run out of other choices
(p. 393)

14. What percent of Brazil's transportation fuel demand was met with ethanol in 1999?
     a) 0.9
     b) 12
     c) 32
     d) 41
     e) 99
(p. 394)

15. Which firm is working to promote competition in the Canadian electricity sector in an effort to stop unnecessary electricity plants?
    a) MANIC
    b) La Grande Research
    c) Energy Probe Research Foundation
    d) Carillon Inc.
    e) Three George

(Pp. 395-396)

16. What was used to enclose the damaged reactor at Chernobyl?
    a) a concrete sarcophagus
    b) an ice cap
    c) cold running water (-10C)
    d) dynamite
    e) asbestos

(p. 397)

17. What Canadian fund was established to help stimulate environmental technologies?
    a) CANDU shares (1998)
    b) Alternative Energy Green Dollars (1999)
    c) Green Power Grants (1996)
    d) Sustainable Development Technology Fund (2000)
    e) Sun & Wind Chapters (2001)

(p. 400)

18. Where is the largest solar converter of water into hydrogen?
    a) Lawrence Livermore, Texas
    b) Tokyo, Japan
    c) El Segundo, California
    d) Black Forest, Germany
    e) Palm Desert, USA

(p. 401)

19. Most countries will NOT meet the Kyoto protocol's CO_2 emissions target by _____.
    a) 2001 to 2005
    b) 2003 to 2007
    c) 2005 to 2011
    d) 2006 to 20011
    e) 2008 to 2012

(p. 408)

CHAPTER 12—WILD SPECIES AND NATURAL SPACES
(pp. 413-448)

STUDENT LEARNING OBJECTIVES

After studying this chapter you should be able to:
- understand the biodiversity concerns relating to Canada's species and spaces
- identify a range of human uses of wild species and natural spaces
- describe the impacts of human activities on wild species and their habitats and environments, including protected areas
- appreciate the complexity and interrelatedness of wild species, habitat, protected areas, and biodiversity issues
- outline Canadian and international responses to the need for protection of wild species and spaces
- discuss challenges to a sustainable future for wild species and protected areas in Canada
- define and discuss all of the KEY TERMS listed below

KEY TERMS (defined at the page number shown and in the glossary)

anthropogenically	422	in situ conservation	434
ecological health	437	intrinsic value	417
ecological integrity	437	organochlorines	427
ecosystem management	438	montane	428
endemic species	418	taxonomic	418
ex situ conservation	439		

NOTES

Introduction
The loss of frogs from geographically dispersed and apparently pristine protected areas suggested one or more global agents might be adversely affecting amphibians.

The importance of biodiversity

Intrinsic value of biodiversity

Intrinsic value	Extrinsic value (research this key work)

Endemic species

Human Activity & Impacts on Canadian Species & Natural Environments

Human activity and biodiversity

Kingdom	Major Subdivision & Common Names (see Table 12-1)
Prokaryotae	
Protistae	
Fungi (Eumycota)	
Plantae	
Animalia	

Habitat alterations due to physical changes

Forestry, Agriculture and Other Human Activities:
Fragmentation:
Chemical Change:
Climate Change:
Habitat Alteration due to Competition from non-native biota:
Habitat Alteration due to Harvesting:
Habitat Alteration due to Toxic contaminants:
Habitat Alteration due to Cumulative agents of change:

Species at risk

The use of wildlife indicators to track toxic contaminants in ecosystems (Box 12-2)

COSEWIC status category, definition & examples of species at risk (Tables 12-3 & 12-4)

Status category	Definition	Example
SC:_____ _____		
T:_____ _____		
E:_____ _____		
XT:_____ _____		
X:_____ _____		
NAR:_____ _____		
DD:_____ _____		

Spaces at risk *(see also Enviro-focus 12 and Box 12-3)*

Responses to environmental impacts and change

International calls to action
Six important documents (see Box 1-1 and Table 1-2)

World Conservation Strategy (1980)	Our Common Future (1987)	Caring for the Earth (1991)	Global Biodiversity Strategy (1992)	Agenda 21 (1992)	Earth Charter (2000)

International Treaties (see also Table 12-5 & 12-6))

In situ conservation

Protected areas (see also Tables 12-6 and 12-7, Figures 12-3 and 12-4, and Box 12-4)

Canada's national parks

Ecological integrity	Ecological health	Ecosystem management

Other protected areas in Canada

Restoration and rehabilitation

Ex situ conservation

Ex situ plant conservation	Ex situ animal conservation

Sustainable use of biological resources

Improving understanding of biodiversity

Canadian law, policy, and practice *(see also Table 12-8)*

Protecting Canadian Species

COSEWIC	Bill C-33

Protecting Canadian Spaces (see Report Card, Box 12-5)

Partnerships for the future

Partnership	Notes
RENEW	
NAWMP	
CPAWS	
CORE	

Future challenges

TRACKING YOUR PROGRESS
Chapter 12—Wild Species and Natural Spaces

Multiple Choice Questions

1. Drost & Fellers (1996) reported the apparent extirpation of this amphibian:
    a) Leopard frog
    b) Golden toad
    c) Gastric breeding frog
    d) Red-legged frog
    e) Bullfrogs
(p. 413)

2. Which of the following is NOT a reason for the world decline of amphibians?
    a) increase in UV-B
    b) radiation resulting from ozone layer depletion
    c) virus found in migratory bird droppings
    d) acid precipitation contamination
    e) pesticides, herbicides and fertilizers
(p. 413)

3. Which organization's goal is to develop a statistically defensible program to monitor the distribution and abundance of amphibians?
    a) CITIES
    b) RENEW
    c) NAAMP
    d) WWF
    e) OFAH
(p. 415)

4. The loss of wildlife species affects traditional lifestyles and reduces the quality of _____.
    a) viewing
    b) country food
    c) habitat
    d) olfactory experiences
    e) dimorphic displays
(p. 417)

5. What is an example of the intrinsic value derived from a peacock?
   a) BBQ food
   b) crafts made from the feathers
   c) petting zoo fees
   d) the colourful show displayed by its fanning tail feathers
   e) none of the above
(p. 417)

6. The cyanobacteria and chloroxybacteria belong to the _____ kingdom.
   a) procaryotae
   b) plantae
   c) animalia
   d) protista
   e) fungi
(p. 419)

7. What fraction of the world's wetlands are found in Canada?
   a) 1/2
   b) 1/3
   c) 1/4
   d) 1/5
   e) 1/6
(p. 421)

8. Which of the following terms does NOT mean non-native?
   a) alien
   b) exotic
   c) endemic
   d) nonindigenous
   e) introduced
(p. 422)

9. An action caused by humans, such as the introduction of lead in nature, is said to be
   _____.
   a) anthropogenic
   b) homocentred
   c) selfworth
   d) endemic force
   e) personal
(p. 422)

10. The loss of _____ can have a "domino effect" on the total functioning of the ecosystem.
     a) humans
     b) eagles
     c) hogs
     d) insects
     e) wolves
(p. 423)

11. Oil exploration near Porcupine caribou herds can cause _____ .
     a) reduced calf production
     b) higher calf survival rates
     c) increased calf production
     d) lower calf survival
     e) both (a) and (d) are correct
(p. 423)

12. Which is the only deer family member that has adapted to life in the Canadian High Arctic?
     a) Bluenose caribou
     b) Peary caribou
     c) Porcupine caribou
     d) Canadian Moose
     e) Black tail
(p. 425)

13. DDE, a compound that inhibits eggshell formation in female birds comes from _____ .
     a) plants
     b) protozoans
     c) insects
     d) birds
     e) DDT
(p. 427)

14. Parks Canada built some highway underpasses for wildlife but _____.
     a) bears like to use them too
     b) bears have cubs in them too
     c) bears hang-out in them too
     d) bears eat their kill in them too
     e) bears do not like using them
(Pp. 428-429)

15. What mode of transport killed 12 species of mammals according to a 1997 Report?
    a) car
    b) bus
    c) plane
    d) train
    e) boat
(p. 430)

16. Why are birds attracted to railway tracks?
    a) They are great to perch on at night.
    b) Birds eat the insects that eat the wood supporting them.
    c) They are cool year round and help cool down the bird after a hard flight.
    d) They feed on mice that eat the various grains that spill from passing trains.
    e) Bird are not attracted to railway tracks.
(p. 430)

17. What acronym did COSEWIC (2000) use to indicate a species that no longer exists?
    a) X
    b) XT
    c) T
    d) DD
    e) SC
(p. 431)

18. Which of the documents below inspired thinking about biodiversity and sustainability?
    a) Our Common Future (1987)
    b) Caring for the Earth (1991)
    c) Earth Charter (2000)
    d) Agenda 21 (1992)
    e) All of the above
(p. 432)

19. What international treaty deals with the protection of wild species?
    a) CITIES
    b) UNEP
    c) WCMC
    d) CBIN
    e) WAPPRIITA
(p. 432)

20. What type of conservation is used to protect a Mountain Gorilla in his homeland?
    a) ex situ conservation
    b) in situ conservation
    c) endemic conservation
    d) intrinsic conservation
    e) extrinsic conservation
(p. 434)

21. The goals of _____ include conservation, education, and recreation.
    a) nature reserves
    b) national parks
    c) protected landscape or seascapes
    d) habitat and species management areas
    e) natural monuments
(p. 435)

22. Which Act provides for the establishment of marine protected areas?
    a) Canadian Oceans Act
    b) Canadian Marine Act
    c) North American Seas Act
    d) North American Marine Charter
    e) Canada's SEA to SEA Act
(p. 438)

23. The goal of _____ is to manage entire ecosystems rather than isolated parts
    of systems.
    a) conservation authorities
    b) wilderness areas
    c) protected seascapes
    d) ecosystem management
    e) nature reserves
(Pp. 438-439)

24. Which of the following reserves have a wide range of objectives including
    monitoring?
    a) Ramsar Sites
    b) Biosphere Reserves
    c) World Heritage Sites
    d) Marine Protected Areas
    e) National Parks
(p. 439)

25. What program is designed to increase populations so species can be returned home?
    a) Safe-Land Inc.
    b) New-Zoo Program
    c) Aqua-Life Program
    d) Species Survival Program
    e) Darwin-Inn Co.
(p. 440)

26. What is the name of the Act that will replace the Endangered Species Protection Act?
    a) Species in Decline Act
    b) Threatened Species Act
    c) Endangered Species Act
    d) Rare Species Convention Act
    e) Bill C-33, Species at Risk Act
(p. 442)

27. The best Endangered Spaces Report card (1998-1999) grade was awarded to _____.
    a) Terrestrial
    b) Ontario
    c) Marine – Great Lakes
    d) Alberta
    e) Quebec
(Pp. 443-444)

28. Which legislation does NOT provide hope for protecting the future of biodiversity?
    a) RENEW
    b) GONE
    c) WAPPRIITA
    d) Migratory Birds Convention Act
    e) Species at Risk Act
(Pp. 444-446)

29. Remember that actions do NOT have to be _____ and _____ to be significant.
    a) given; taken
    b) forwarded; delivered
    c) front; forward
    d) silver; gold
    e) big; expensive
(p. 446)

CHAPTER 13—LIFESTYLE CHOICES AND
SUSTAINABLE COMMUNITIES
(pp. 450-483)

STUDENT LEARNING OBJECTIVES

After studying this chapter you should be able to:
- understand the main issues and concerns relating to urbanization and ecosystems in Canada.
- identify the effects of urbanization on the environment around us.
- appreciate the complexity and interrelatedness of issues relating to urbanization and the environment.
- appreciate the effects our lifestyle choices have on our environment.
- understand the types of efforts Canadians have made toward more sustainable communities.
- define and discuss all of the KEY TERMS listed below.

KEY TERMS (defined at the page number shown and in the glossary)

NOTES

Introduction
We often seem to forget that our cities are part of the ecosystem.

Urban Environmental Conditions and Trends

Atmosphere and Climate

Microclimate (see also Box 13-1)	Air quality (see Box 13-2 & Figure 13-1)

Noise *(see also Box 13-3)*

Water
Cities and the hydrological cycle

Water supply and water quality

Water use and wastewater treatment (see also Table 13-1, Figures 13-2a and 13-2b)

Water and recreation

Energy

Sustainable housing (see also Figure 13-3, Table 13-2)	Depending less on our cars - Sustainable transportation

Materials use

Solid waste (see also Figure 13-4, Box 13-4)	Land contamination

Urbanization of land

Green space in the City (see also Box 13-5)	Loss of agricultural land (see also Box 13-6)

Towards sustainable communities
What if cities did not grow?

Cities and sustainability
How much land does the average Canadian need in order to provide the resources consumed?

Making Canadian cities more sustainable

Urban Form (see also Table 13-3)	
Conservation: 1. Water conservation 2. Energy conservation 3. Conservation of materials	
Conservation of Ecosystems and Natural Features (see also Table 13-4)	
Reduction of Environmental Impacts: 1. Air quality 2. Water quality 3. Waste management	
Transportation	
Planning	

Progress Toward Urban Sustainability?

TRACKING YOUR PROGRESS
Chapter 13—Lifestyle Choices and Sustainable Communities

Multiple Choice Questions

1. What percent of Canadians live in communities with over 1000 people?
  a) 75
  b) 98
  c) 84
  d) 66
  e) 42
(p. 451)

2. On the HDI, Canada ranked _____ among all countries in the world (2000).
  a) 5
  b) 4
  c) 3
  d) 2
  e) 1
(p. 451)

3. Buildings and streets, reduction of wind speed, and atmospheric pollutants can create a

 _____
  a) city VOC
  b) heat island
  c) macroclimate
  d) cooling effect
  e) huge wind factor
(p. 452)

4. What common pollutant does NAPS NOT measure?
  a) sulphur dioxide
  b) nitrogen dioxide
  c) nitrous oxide
  d) carbon monoxide
  e) suspended particles
(p. 452)

5. What is one of the most common occupational diseases in industry today?
  a) sight impairment
  b) hearing loss
  c) stress
  d) cancer
  e) alcoholism
(p. 455)

6. In 2000, the Drinking Water Protection Regulation came into effect in which province?
   a) Alberta
   b) Manitoba
   c) Quebec
   d) Ontario
   e) Nova Scotia

(p. 456)

7. When _____, they used 60 per cent more than Edmonton residents on water meters.
   a) Windsor residents paid for bottled water
   b) Calgary residents paid a flat rate for water
   c) Whitehorse residents paid a water meter rate for potable water
   d) Edmonton residents paid a fee for well water
   e) Lindsay residents paid a private business to boil their water

(p. 457)

8. Where is thirty percent of all sewage generated released untreated into the environment?
   a) British Columbia
   b) Alberta
   c) Ontario
   d) Quebec
   e) Nova Scotia

(p. 458)

9. What type of pollution do powerboats cause?
   a) noise
   b) water
   c) air
   d) all of the above
   e) none of the above

(p. 460)

10. The oven associated with the ASH house is found in the _____ facing the sun.
   a) front porch
   b) kitchen
   c) dinning room
   d) foyer
   e) garage

(p. 461)

11. Walls made out of _____ have an insulation value approaching R-50.
    a) tires
    b) wood
    c) straw bales
    d) bricks
    e) stone
(p. 461)

12. What constitutes the bulk of municipal and construction waste streams (1992)?
    a) wood and yard wastes
    b) plastics and glass
    c) metals and gypsum
    d) asphalt, paper and paperboard
    e) paper and concrete
(Pp. 465-466)

13. Apart from its important social benefits, most urban outdoor recreation space has little
    _____.
    a) conservation value
    b) ecological value
    c) environmental value
    d) none of the above
    e) all of the above
(p. 468)

14. What is the urban shadow effect?
    a) agricultural impacts that extend into urban areas
    b) urban impacts that extend over large areas and cause declines in agriculture
    c) agricultural impacts that extend into wildlife areas
    d) urban impacts that extend into wildlife areas and cause declines in rural life
    e) rural impacts that extend into urban zones and cause agricultural declines
(p. 470)

15. What occurs when recreational facilities and residences are developed near agricultural regions?
    a) macroclimatic effects
    b) urban shadow effects
    c) agricultural shadow effects
    d) microclimatic effect
    e) no effect
(p. 470)

16. Human society consumes 30% more of the earth's natural capital than is generated:
    a) every five years
    b) each year
    c) every generation
    d) every 10 years
    e) every 100 years
(p. 472)

17. This is NOT (but perhaps should be) a principle of sustainable cities:
    a) protect wetlands and their riparian zones at all costs
    b) promote social equity
    c) maintain or enhance ecological integrity
    d) encourage democratic participation of all citizens
    e) meaningful work and livelihood for all citizens
(p. 472)

18. In the future, which of the following characteristics will "sustainable cities" have?
    a) rapid transit
    b) corridors of open space
    c) mixed land use pattern
    d) moderate- to high-density housing
    e) all of the above
(Pp. 473-474)

19. What is a good method for people and communities to augment municipal water supply?
    a) dig private water wells
    b) buy bottled water
    c) collect rainwater in barrels
    d) tap into natural springs
    e) gather water from artificial creeks and ponds
(p. 475)

20. What notion is the ecosystem approach based on?
    a) culture is not important
    b) everything is connected to everything else
    c) living species are most important
    d) air, land, and water are the main focus
    e) encompass physical relationships above all
(p. 476)

21. Which Internet address gives you access to the Trans Canada Trail page?
    a) http://www.trail4u.ca/
    b) http://www.transtrail.ca/
    c) http://www.TCT.ca/
    d) http://www.tctrail.ca/
    e) http://www.hikecanada.ca/
(p. 477)

22. What is the source for "Actions to achieve more sustainable urban transportation"?
    a) The State of Canada's Future, 2000
    b) The State of Canada's Environment, 1996
    c) Supply and Services Ontario, 1999
    d) Supply and Services Alberta, 2001
    e) all of the above
(p. 478)

23. What is an alternative to the GDP?
    a) GDE
    b) GDI
    c) GDH
    d) GDG
    e) GDC
(Pp. 479-480)

24. "Good intentions and expressions of concern are NOT enough. Protecting the environment is the most serious challenge…today…get involved…every day."
    a) Dianne Draper
    b) John Doe
    c) Roberta Bondar
    d) David Suzuki
    e) Jean Chrétien
(p. 480)

CHAPTER 14—MEETING ENVIRONMENTAL CHALLENGES
(pp. 484-508)

STUDENT LEARNING OBJECTIVES

After studying this chapter you should be able to:
- appreciate the general trends in environmental issues and sectors
- outline the broad range of Canadian efforts to safeguard our environment
- identify those challenges that remain in our quest for sustainability
- appreciate some of the Earth-sustaining actions we can take
- define and discuss all of the KEY TERMS listed below

KEY TERMS (defined at the page number shown and in the glossary)
Review Glossary

NOTES

Introduction
Education is central to the making of appropriate and sustainable decisions.

Progress in Safeguarding Canada's Environment

Air quality Issues

Water quality issues

Biological diversity *(see also Figure 14-1)*

Climate change *(see also Figure 14-2).*

Sector industries

Agriculture	
Forests	
Minerals and metals	
Energy	
Fisheries	

Regulatory efforts to safeguard our environment

Canada

Atlantic Canada	Central Canada	Western Canada & the North

ENGO Actions to Safeguard our Environment

Challenges for the Future

Resource management	
Conservation	
Waste reduction (Box 14-1)	
Urban centres and transportation	
Pollution control (Box 14-2)	
Cleanup of past environmental problems (Enviro-Focus 14)	
Changes in decision-making processes (Box 14-3)	

The importance of individuals

Critical assessment

Use of appropriate technology:

Information and education:

Simplicity

What legacy do we want to leave for succeeding generations?

TRACKING YOUR PROGRESS
Chapter 14—Meeting Environmental Challenges

Multiple Choice Questions

1. Not only do we need to think globally and act locally, but also we need to _____.
    a) recognize the fundamental issues involved in achieving sustainability
    b) be cognizant of the impacts our actions may have on others
    c) work together to achieve goals
    d) all of the above
    e) some of the above
(p. 485)

2. In what year was lead phased out as a gasoline additive for road vehicles?
    a) 1962
    b) 1990
    c) 2001
    d) set for 2008
    e) there is no plan to phase out lead from gas as it is most essential
(p. 486)

3. When is particular matter less than or equal to 2.5 microns in diameter to be phased out?
    a) already in place since 1992
    b) 2002
    c) 2008
    d) 2010
    e) 2014
(p. 486)

4. What is central to the making of appropriate and sustainable decisions?
    a) government
    b) municipalities
    c) the Windsor-Quebec corridor
    d) COSEWIC
    e) education
(p. 486)

5. Which Act will include Aboriginal peoples, farmers, scientists, and industry?
    a) DDT
    b) ENGO
    c) COSEWIC
    d) SARA
    e) PCB
(p. 487)

6. Donations of ecologically sensitive land are encouraged through _____.
    a) telephone marketers
    b) federal tax law changes
    c) inheritance gifts
    d) local police funds
    e) Ducks Unlimited
(p. 487)

7. Why is the reduction of fossil fuel use difficult in Canada?
    a) our large land mass
    b) cold climate
    c) increasing population
    d) some of the above
    e) all of the above
(p. 488)

8. Trends indicate that the growth rate of greenhouse gas emissions is _____.
    a) slowing in Canada
    b) increasing in Canada
    c) decreasing in Canada
    d) stable in Canada
    e) augmenting in Canada, far beyond those of all other countries combined
(p. 488)

9. Efforts to maintain and improve water quality have NOT included the management of

    _____.
    a) soils
    b) nutrients
    c) wetlands
    d) manure
    e) pesticides
(p. 489)

10. R-2000 homes need to be _____.
    a) encouraged
    b) invented
    c) dismantled
    d) larger
    e) smaller
(Pp. 489-490)

11. On what principle is the Ocean Act built?
   a) When two species are competing for the same resources, one must change.
   b) Matter is neither created nor destroyed.
   c) Everything in the natural world is connected.
   d) Lack of full scientific certainty shall not be used as a reason for postponing action.
   e) An insistence that we fully understand a problem before we take action.
(p. 490)

12. In which province was $6 million allocated to groundwater monitoring in 2000?
   a) Alberta
   b) Ontario
   c) BC
   d) PEI
   e) Nova Scotia
(p. 491)

13. Environmental protection challenges in Saskatchewan did NOT include _____.
   a) underground storage tanks
   b) monitoring and managing acid deposition
   c) climate change issues
   d) addressing contaminated sites
   e) assessing on-land methane discharge levels from the bovine industries
(p. 494)

14. The Government of Alberta has invoked an ecological approach to better manage _____.
   a) oil industries
   b) recreation sport use
   c) mining
   d) forests sustainably
   e) urban development
(p. 494)

15. Which province has reached its commitment of protecting 12% of its lands & waters?
   a) British Columbia - under the Canadian Wilderness Charter
   b) New Brunswick - under the Canadian Wilderness Charter
   c) Manitoba - under the Canadian Wilderness Charter
   d) Newfoundland - under the Canadian Wilderness Charter
   e) Québec - under the Canadian Wilderness Charter
(Pp. 494-495)

16. How many planets would it take for everyone to equal Canadians in resource use?
    a) one planet Earth
    b) two planet Earth
    c) three planet Earth
    d) four planet Earth
    e) five planet Earth
(Pp. 496–497)

17. The term "live within our means" was made evident by the _____.
    a) extinction of the grizzly bears in the Rocky mountain foothills of Alberta
    b) collapse of the northern cod stocks
    c) explosion of immigrants entering Bangladesh in 2000
    d) landings on planet Mars these last years
    e) lack of fuel in Canadian homes last winter
(p. 498)

18. How many two-litre plastic pop bottles are needed to make an average size jacket?
    a) 7
    b) 12
    c) 25
    d) 46
    e) 78
(p. 498)

19. Where does the water, food, energy, and minerals that flow into our cities come from?
    a) industry
    b) grocery stores
    c) other countries
    d) surrounding ecosystems
    e) nowhere
(p. 499)

20. Although government resources have declined, the increasing importance attached to _____.
    a) environmental auditing has subsided
    b) industrial ecology has flourished
    c) self-regulation has taken-off in a positive note
    d) protecting ecologically sensitive areas in urban landscapes has grown
    e) re-developing the transportation infrastructure has stabilized
(p. 500)

21. Who expects public and private institutions to achieve environmental quality?
    a) deep ecologists
    b) CEPA
    c) environmentalist
    d) resource managers
    e) the Canadian public
(p. 500)

22. The Toxic Substances Management does NOT make it easy to _____.
    a) identify pollution sources
    b) track pollutant releases
    c) more effectively promote management of polluting processes
    d) do all of the above
    e) do none of the above
(p. 501)

23. Environmental auditing allows remedial or preventive actions to be taken before_____.
    a) civil damages or noncompliance with regulations result in jail time
    b) civil damages or noncompliance with regulations result in mitigation processes
    c) civil damages or noncompliance with regulations result in high costs
    d) industry is allowed to devise ways to decrease negative impacts
    e) civil damages or noncompliance with regulations result in more legislation
(p. 501)

24. Sustainability approaches for Earth-sustaining actions do NOT include _____.
    a) ecosystem management
    b) mitigating processes
    c) collaboration and co-operation among individuals, private sector and governments
    d) an integrated approach
    e) technology sharing
(Pp. 504-506)